RSME Springer Series

Volume 16

As of 2015, RSME - Real Sociedad Matemática Española - and Springer cooperate in order to publish works by authors and volume editors under the auspices of a co-branded series of publications including advanced textbooks, Lecture Notes, collections of surveys resulting from international workshops and Summer Schools, SpringerBriefs, monographs as well as contributed volumes and conference proceedings. The works in the series are written in English only, aiming to offer high level research results in the fields of pure and applied mathematics to a global readership of students, researchers, professionals, and policymakers.

Angel Ferrandez Izquierdo

Helices

Real Sociedad Matemática
Española

Springer

Angel Ferrandez Izquierdo
Department of Mathematics
University of Murcia
Murcia, Murcia, Spain

ISSN 2509-8888 ISSN 2509-8896 (electronic)
RSME Springer Series
ISBN 978-3-032-09662-3 ISBN 978-3-032-09663-0 (eBook)
https://doi.org/10.1007/978-3-032-09663-0

This Springer imprint is published by the registered company Springer Nature Switzerland AG
The registered company address is: Gewerbestrasse 11, 6330 Cham, Switzerland

To my friend Manolo, the soul of this book.

To Marisol, Marta, Mar & Mar, and Héctor & Héctor, the engines of my life.

And to the memory of Antonio M. Naveira and Óscar J. Garay.

Preface

The paramount aim of scientific research is the knowledge of Nature and the laws that govern it. Kant's categorical assertion that the criterion of true science lay in its relation to mathematics may not hold today, or at least would be severely criticized. D'Arcy Thompson, in his excellent book *On growth and form* [101], which we will address later, reminds us of Sir John Herschel's words: "Numerical precision is the very soul of science, and its attainment affords the best, perhaps, the only criterion of the truth of theories and the correctness of experiments," and quotes again to Kant when he warned that "it was Nature herself, and not the mathematician, who brings mathematics into natural philosophy."

The choice of helices to write this book has not been random. It is enough to see our recent research career to see that they have been a recurrent research subject. We would like to highlight, if that were possible, this mathematical structure with great preponderance in Nature and even in our daily lives. Moreover, we intend to collect many somewhat scattered results and put them in a unique context using helices as leitmotiv.

In academia, the best known helical configuration is that associated with the DNA molecule. After its discovery by Rosalind Franklin, James D. Watson, and Francis H. C. Crick (see [105]), a myriad of articles on that famous double helix has flooded the scientific literature. Everybody is aware of the ubiquity of helices in Nature and admires their intrinsic beauty, but it seems that people are not interested in knowing the omnipresence of helices in the real world. People know the value of a corkscrew, or a spiral staircase, but it is not so easy to understand the reasons of supercoiling of the strands making up the DNA molecule.

The reader should know that we have not attempted an exhaustive work on helices. We have only collected a tiny fraction of the literature on the subject, which essentially interested us from the purely geometric point of view. Suffice it to say that on the occasion of the celebration of the half century of the discovery of the double helix structure of the DNA molecule, it became revealed that annually were published around 2,000 papers on that structure.

We wanted to pay special attention to the helical configurations that are plentiful in Nature and we usually find in our close environment. Based on the seminal paper

General helices and a theorem of Lancret, by Prof. M. Barros (see [6]), all those curves, which we call **Lancret, or generalized, helices**, lie in the three-dimensional Euclidean space $\mathbb{R}^3$ and its tangent vector field makes a constant angle with a fixed direction. This definition will be reformulated depending on the context. Thus, as the reader will soon realize, the first chapter is the core of this work, where we exhibit a series of results where the helices play a special role because of the beauty of the solutions they provide and the originality of unsuspected achievements.

Without getting into great speculations, we soon realize that Nature is governed by a principle of minimum energy, that is, any natural process occurs at minimum cost. With that idea in mind, and thinking in a supercoiled DNA molecule, we are predisposed to think that this phenomenon reflects the need to place the greatest amount of matter in the least possible space. Or even, that the arrangement in a double supercoiled helix is the optimal organization to facilitate the mechanism of replication. That said, we should think of some kind of energy functional, defined on a suitable space of curves, and determine its critical points.

The question of closing a helix is of supreme importance, and it will be addressed avoiding $\mathbb{R}^3$. Indeed, the electron microscopy offers photographs of closed helical configurations, i.e., closed Lancret helices, that could consider winded around a donut, which is a revolution surface mathematically known as a torus. But now, what is the axis of the helix?

We intended deliberately that the level of this work is suitable for a Master of Science, linking with the expertise of a Bachelor in Mathematics or Physics and leaving an open door to research world. In addition, we have tried to make a self-contained book, for which we have included several appendices. And, wishing that our eyes see beyond, we have deployed a wide and unsuspected series of helices in geometries where lightlike vectors play a key role.

To assist the reader, it should be noted that the work is organized in four parts. We first consider the Riemannian cases and then the Lorentzian analogues. We also consider some related questions concerning variational problems on curves and surfaces, and dynamic of particles. Finally, the last one contains two appendices to complete and extend some details that we believe will be useful for people interested in the topics discussed here.

My heartfelt thanks to my friend M. Barros for his infinite generosity in always sharing with me his profound mathematical knowledge. Without his help this book would never have seen the light of day.

This book has been written thanks to the nice atmosphere created around a coffee cup, at 11 o'clock every day, by my colleagues Profs. Pascual Lucas, Luis J. Alías, Miguel A. Meroño, Miguel A. Javaloyes, Ángel Giménez, José A. Pastor, and Ildefonso Castro. A very special thanks to Ma Ángeles H. Cifre, Jesús Yepes, Lidia Gordo, and Francisco Carreras for his technical support in bringing this book to its final form.

I would like to express my deepest gratitude to the two reviewers for their patience, interest, and generosity in reading my manuscript. They have gone into it with the necessary depth to show me a number of details that I had missed. Indeed, they have reviewed the manuscript from a point of view farther away than mine, that is, as astronauts observing Earth from the International Space Station. As a result, they have given me a number of invaluable suggestions that have contributed greatly to the improvement of the final version of the book.

Murcia, Spain Angel Ferrandez Izquierdo

Introduction

The book by T. A. Cook, *The Curves of Life* (see [33]), contains an stimulating subtitle: "An account of spiral formations and their application to growth in nature, to science and to art." Cook distinguished between spiral and helix, the former when coiling in a plane, the latter when coiling around a central axis. In general, we will talk about helical configurations. That excellent book shows a lot of examples of helical configurations such as shells of mollusks, horns of sheep, goats, and antelopes; bones, feathers, arrangements of teeth in sharks and leaves. In the Preface, we read "In shells, in plants, in the bodily structures of men and animals, the spiral formation is certainly a common factor in a multitude of phenomena apparently widely different ... The Formula for Growth now suggested in this book is here called the Φ spiral, or Spiral of Pheidias." Today, we will say the golden ratio.

Shortly after, another major work, *On growth and form* [101], his author, D'Arcy W. Thompson, paid special attention to the form as a expression of growth. As Thompson was strongly convinced that "numerical precision is the very soul of science," his book was decorated with a remarkable mathematical content.

As for highlighting helical configurations in Nature, two works deserve our special attention. In the first one, by J. Galloway [47], you can read "Helices appear at every anatomical level across the nine (or so) orders of magnitude that span the range of size between molecules and the biggest organisms. They provide solutions to any number of the challenges of growth and form, structure and function including significantly movement, that evolution has thrown up. This essay explores the helix both as an abstract mathematical idea, with its stark elegance, simplicity and economy, and the 'real' helical structures that contribute to the richness and complexity of the living world, and the relationship between them. Helical structures are so pervasive that the helix can perhaps be regarded as providing a unifying and even necessary structural principle for life. The helical idea goes a long way to explain why life at its most fundamental level of genes and proteins depends on two classes of small enantiomeric molecules, amino acids and nucleotides, significantly the molecules in each of the two classes all possessing the same hand. Life's apparent requirement for helical symmetry at the deep molecular level forces the uniformity."

The second one, by N. Chouaieb et al. [32], states that "Helices are among the simplest shapes that are observed in the filamentary and molecular structures of nature. The local mechanical properties of such structures are often modeled by a uniform elastic potential energy dependent on bending and twist."

We are mainly interested in the geometry of helices, so we are inspired by the seminal paper by M. Barros [6]. A curve of constant slope or general helix in Euclidean space $\mathbb{R}^3$ is defined by the property that the tangent makes a constant angle with a fixed straight line (the axis of the general helix). A classical result stated by M. A. Lancret in 1802 and first proved by B. de Saint Venant in 1845 (see [94] for details) is: *A necessary and sufficient condition that a curve be a general helix is that the ratio of curvature to torsion be constant.*

In addition to the above overviews, we will offer exciting new helical structures with physical and biological interest. For instance, we will talk about magnetic curves and chain proteins.

We have all found, and surely marveled with, some helical structure. The shell of a snail, a climbing plant, a corkscrew, the tail of a seahorse, the horns of many families of goats, or many types or spiral galaxies are examples at first sight.

Some natural helical structures

At the microscopic level it is known that the double-stranded DNA is a double helix. The principal secondary structures in proteins are α-helices and β-sheets, which are sheets of helices. The fibrous protein α-keratin is a double α-helix; collagen is a triple helix. The cytoskeletal filaments—actin filaments, microtubules, and intermediate filaments—are helical assemblies of subunits. Helices occur in the capsids of viruses.

In some ancient civilizations, the idea of helix appears linked to divine belief of "lack of beginning and end." For instance, in Mayan culture our current number zero was represented by a small helix.

As a general rule, by a *helicoidal configuration* we will mean any structure based on a helix. For instance, a screw or a spiral staircase. Looking at these examples, one realizes that those configurations, both being based on a helix, present a substantial difference, since the former is wound on a cone, while the latter does so on a right circular cylinder, i.e., a cylinder built with straight lines which are leaning orthogonally on a circle.

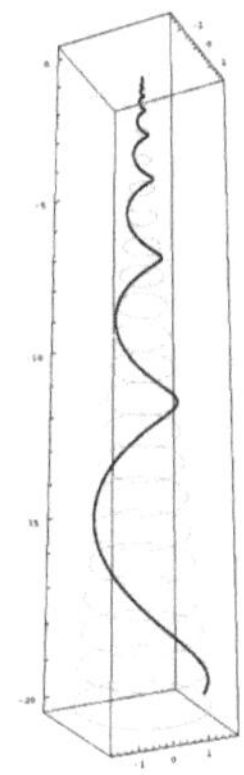

A conical helix

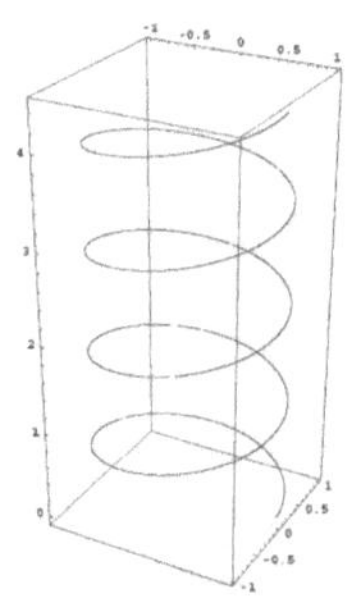

A circular helix

The difference shown is fundamental, because if we had to bind up an injured arm, we would realize that the movement of the bandage would be helical, but the surface on which it would be rolled up would be neither conical nor cylindrical. Then it appears a new helical structure. The same thing happens in a protein or in a DNA molecule, where we find out helical configurations formed by helices wounding around other helices.

It is thus that Nature is bountiful in structures based on helices. In fact, the helical configurations are extremely important, because molecules of life, as proteins and DNA, are forming such structures.

But then, what is the ultimate reason for the proliferation of helical configurations in Nature? For instance, one would think that the double helix structure of a DNA molecule responds to the need of making the best use regarding the available limited space. At the macroscopic level, to open a bottle, the corkscrew is smoothly inserted into the cork to remove it, that is, following a helical path we have achieved a substantial energy saving when drilling the cap.

However, there is more, because a well-known result in the world of Biology, which could be stated as the Pauling Theorem ([85], [83], and [84]) states that "identical objects, regularly assembled, form a helix." Furthermore, following Cahill

[27] "this result is the principal motif of nucleic acids, proteins, and viral capsids acquire their helical configurations." In that paper Cahill gives the first known proof of the Pauling Theorem.

As we have just said, in most cases, and for the sake of simplicity, helical configurations are often identified with helices winding on right circular cylinders, which is far from what happens in Nature. Indeed, nobody can believe that squirrels chasing one another up and around tree trunks follow a path of a circular helix. First because the transversal section of a tree trunk is not circular, but also because the axis of a three trunk is not exactly a straight line.

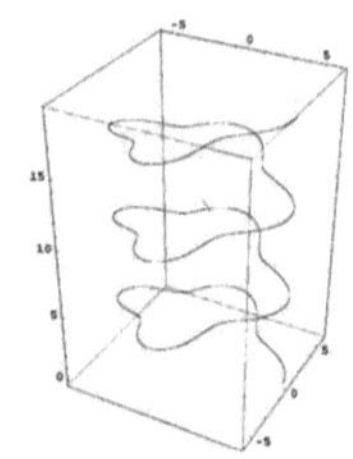

A starlike helix

We see that if we consider any plane curve, not even closed, and draw perpendicular lines on it, we will construct a cylinder that is more general than the circular one. If a rope is winded around the cylinder, we get a helix. Furthermore, for that kind of helices, we observe two interesting facts:

(1) They have an axis, that is, a fixed curve—in particular, a straight line—which makes a constant angle with the tangent vector of the helix; and
(2) They never close up, because its tangent vector always has a positive component, precisely in the axis direction.

Now, if that rope is winded around a donut, then we could expect a closed helicoidal configuration. And that is precisely what happens in a larger number of circular proteins, in bacteria, plants, and animals whose components are forming closed helices. In this case, the axis is a closed curve.

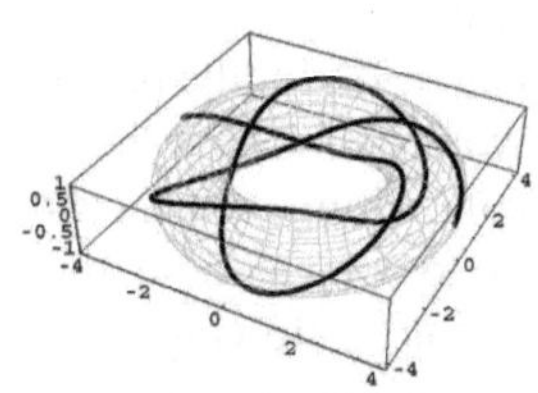

A closed helix on a torus

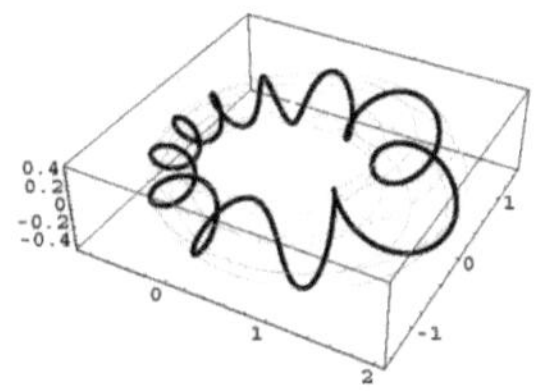

A closed helix on a Dupin cyclide

Furthermore, the theory of helices falls into place in a broader scenario. For instance, helicoids or surfaces with helicoidal symmetry basically inherit all their properties from their generating curves, the helices, and also arise in many contexts. As just one example, the minimal helicoid has recently been suggested to model how the peripheral stacks of sheets of the rough endoplasmic reticulum are connected (see [51] and [98]).

Finally, we have exhibited a close relation between the theory of Lancret helices and the variational derivatives of their speed, curvature, and torsion. More precisely, Killing vector fields along curves are essential to understand Lancret helices. This leads to, as already pointed out throughout the book, a close relation with variational problems for curves whose energy densities depend on the curvatures. There are many sections on the book where this has been explained and developed.

Contents

Chapter 1
Lancret Helices in $\mathbb{R}^3$

1.1 Circular Helices

It is well known that a straight line and a circle, of a certain radius r, are both plane curves having constant curvature. The former with curvature zero and the latter with curvature $\pm 1/r$, the sign depending on the chosen orientation. The converse also holds. Let $\gamma(s)$ be a plane curve with constant curvature κ_o. If $\kappa_o = 0$ then $\gamma''(s) = T'(s) = 0$ and the curve is obviously a straight line. However, if $\kappa_o \neq 0$, we can differentiate in the curve $\beta(s) = \gamma(s) + \frac{1}{\kappa_o} N(s)$ to find

$$\frac{d}{ds}(\beta(s)) = T(s) - \frac{1}{\kappa_o}\kappa_o T(s) = 0,$$

showing that is a constant function, in other words, $\gamma(s)+\dfrac{1}{\kappa_o}N(s) = p_o$. Therefore,

$$|\gamma(s) - p_o|^2 = \frac{1}{\kappa_o^2},$$

which means that is a circle centered at p_o with radius $1/\kappa_o$.

We wish to find all curves in $\mathbb{R}^3$ with curvature and torsion being both constant. To do that, the above experiment about straight line and circle, up to coordinate changes, can be seen as follows:

(i) Given any real number $b \neq 0$, let us consider the one-parametric group of translations

$$\mathcal{T} = \{\mathrm{T}_{\vec{v}_t} : \vec{v}_t = (0, 0, b\,t), \quad t \in \mathbb{R}\}.$$

This acts on $\mathbb{R}^3$ giving orbits which are straight lines parallel to the z-axis.

A. Ferrandez Izquierdo, *Helices*, RSME Springer Series 16,
https://doi.org/10.1007/978-3-032-09663-0_1

(ii) We can also consider the following one-parametric group of rotations

$$\mathcal{R} = \left\{ \mathrm{R}_t = \begin{pmatrix} \cos t & \sin t & 0 \\ -\sin t & \cos t & 0 \\ 0 & 0 & 1 \end{pmatrix} : t \in \mathbb{R} \right\}.$$

This group acts on $\mathbb{R}^3$ whose orbits are circles lying in planes orthogonal to the z-axis and centered at this axis.

It is clear that any straight line in $\mathbb{R}^3$ can be seen as an orbit of the above translation group on $\mathbb{R}^3$. Similarly, any circle can be seen as an orbit of the above rotation group on $\mathbb{R}^3$. Those facts are useful to justify geometrically the constancy of curvature in both cases. Furthermore, we are on the right way to get new hints. For instance, we look for the orbits of the helicoidal motion group $\mathcal{H} = \{\mathrm{H}_t = \mathrm{T}_t \cdot \mathrm{R}_t : t \in \mathbb{R}\}$ acting on $\mathbb{R}^3$.

Certainly, from its very definition, the curvature and torsion of those curves are both constant. Taking the point $p_o = (a, 0, 0)$, when $a > 0$, the corresponding orbit is the following parameterized circular helix

$$\gamma(t) = (a \cos t, a \sin t, bt),$$

where a is called the radius and b the pitch.

Writing $s = ct = \sqrt{a^2 + b^2}\, t$, we take the bi-parametric family of circular helices given by

$$\gamma_{ab}(s) = \left(a \cos \frac{s}{c}, a \sin \frac{s}{c}, \frac{b}{c} s \right),$$

all of them living in a cylinder of radius a whose generatrices are parallel to the z-axis. Actually, they wound around that cylinder in such a way that its z-coordinate changes linearly and increases or decreases, according to $b > 0$ or $b > 0$, respectively. Then we have got a bi-parametric family of curves with curvature and torsion being both constant.

To show that, up to rigid motions in $\mathbb{R}^3$, they are unique, we will compute the curvatures in terms of the parameters a, b. The unit tangent is given by

$$T(s) = \gamma'_{ab}(s) = \left(-\frac{a}{c} \sin \frac{s}{c}, \frac{a}{c} \cos \frac{s}{c}, \frac{b}{c} \right), \qquad c = \sqrt{a^2 + b^2},$$

and the acceleration vector field by

$$T'(s) = \gamma''_{ab}(s) = \left(-\frac{a}{c^2} \cos \frac{s}{c}, -\frac{a}{c^2} \sin \frac{s}{c}, 0 \right), \qquad c = \sqrt{a^2 + b^2}.$$

By using the Frenet equations, the constant curvature writes as follows

$$\kappa_{ab} = \frac{a}{a^2 + b^2}.$$

The unit normal is

$$N(s) = \left(-\cos\frac{s}{c}, -\sin\frac{s}{c}, 0\right),$$

and so the binormal is given by

$$B(s) = T(s) \wedge N(s) = \left(\frac{b}{c}\sin\frac{s}{c}, -\frac{b}{c}\cos\frac{s}{c}, \frac{a}{c}\right).$$

From here and using again the Frenet equations, we deduce that

$$B'(s) = \left(\frac{b}{c^2}\cos\frac{s}{c}, \frac{b}{c^2}\sin\frac{s}{c}, 0\right).$$

Then we compute the constant torsion by

$$\tau_{ab} = \frac{b}{a^2 + b^2}, \qquad \text{constant.}$$

Observe that the main ingredients (radius and pitch) defining a circular helix can be recovered. It is easy to see that

$$a = \frac{\kappa_{ab}}{\kappa_{ab}^2 + \tau_{ab}^2}, \qquad b = \frac{\tau_{ab}}{\kappa_{ab}^2 + \tau_{ab}^2}.$$

We have interesting consequences. Indeed, from the fundamental theorem of space curves, we deduce that *any curve with non-zero constant curvature and torsion is congruent to a circular helix in the above bi-parametric family*. As usually, we will call circular helix to any curve with constant curvature and torsion. The family of circular helices certainly includes two extremal cases: straight lines and circles, the former seen as zero radius circular helices, the latter as zero pitch ones.

1.2 Lancret Helices

An interesting property of circular helices is that they make a constant angle with some straight line. Indeed, with the above notation, we see that

$$\langle T(s), \vec{e}_3 \rangle = \frac{b}{c}.$$

It could be a good idea to use this property to characterize circular helices. Assume now that $\gamma(s)$ is a curve in $\mathbb{R}^3$ making a constant angle with a fixed direction, which will be called the *axis* of the curve, and take a unit vector $\vec{v}$ on the axis. Let $\gamma(s)$ be parameterized by its arc length, so its unit tangent is $T(s) = \gamma'(s)$. Let θ be the constant angle between the curve and its axis. Therefore we have

$$\langle T, \vec{v} \rangle = \cos\theta.$$

Obviously, we can consider that $\theta \neq 0$, otherwise $\gamma(s)$ is a straight line parallel to the axis. Now we define the following vector field along $\gamma(s)$

$$D(s) = T(s) - \cos\theta\, \vec{v},$$

which lies in the plane $\mathrm{P} = \langle \vec{v} \rangle^{\perp}$ normal to the axis. In that plane we take the $\pi/2$ rotation J, positively oriented according to the induced rotation by that of $\mathbb{R}^3$ in P.

Then we can define the unit normal vector field along $\gamma(s)$ by

$$N(s) = \frac{1}{\sin\theta} J(D(s)),$$

which allows us to consider the Frenet trihedron $\{T, N, B = T \times N\}(s)$ along $\gamma(s)$ without resorting to the acceleration vector field of the curve. As $\langle T'(s), \vec{v} \rangle = 0$, we see that $T'(s) \in \mathrm{P}$. On the other hand, as $||T(s)|| = 1$, one has $\langle T'(s), T(s) \rangle = 0$. Then we get $\langle T'(s), D(s) \rangle = 0$, showing that the acceleration vector field $T'(s) = \gamma''(s)$ and the principal normal $N(s)$ are collinear, which yields the first Frenet equation

$$T'(s) = \kappa(s)\, N(s).$$

The curvature function is given by

$$\kappa(s) = \langle T'(s), N(s) \rangle.$$

It is worth noting that, like in plane curves, the curvature function of the curves making a constant angle with a fixed direction is not signed, that is, it can change sign from one point to another.

Now we differentiate in the equation $\langle N, \vec{v} \rangle = 0$ (any rectifying plane of the curve contains the axis) and use the second Frenet equation to find that

$$\langle -\kappa\, T + \tau B, \vec{v} \rangle = 0.$$

As $\vec{v} = \cos\theta\, T + \sin\theta\, B$, we get

$$\frac{\tau}{\kappa} = \cot\theta.$$

The converse also holds and so that property characterize a sort of "helices" more than circular ones. Indeed, if $\frac{\tau}{\kappa} = \cot\theta$, then

$$(\cos\theta\ T + \sin\theta\ B)' = (\kappa\cos\theta - \tau\sin\theta)N = 0,$$

which implies that $(\cos\theta\ T + \sin\theta\ B) = \vec{v}$ is a fixed vector making a constant angle θ with $\gamma(s)$.

As a consequence, the property of "making a constant angle with a fixed direction" characterizes not only circular helices, but also the curves whose ratio between its curvature and torsion is a constant. From now on, such a curves will be called **Lancret, or generalized, helices** (see [66]).

1.3 A Geometric Vision of Lancret Helices

A right circular cylinder C in $\mathbb{R}^3$, of radius r and generatrices parallel to the z-axis, is defined by the equation $x^2+y^2 = r^2$. It is well known that C can be parameterized in terms of a circle $\alpha(s) = (r\ \cos\frac{s}{r}, \sin\frac{s}{r}, 0)$, of radius r centered at the origin, in the plane $z = 0$ and the vector $\vec{e}_3 = (0, 0, 1)$ as follows

$$\mathrm{C} = \{\alpha(s) + z\,\vec{e}_3\ :\ s, z \in \mathbb{R}\}.$$

That gives the map

$$\mathbf{X} : \mathbb{R}^2 \to \mathrm{C}, \qquad \mathbf{X}(s, z) = \alpha(s) + z\,\vec{e}_3.$$

We know that the helices winding around C are obtained as images by that map of straight lines passing through the origin.

Observe that the above situation can be easily generalized starting from any plane curve, in other words, we are going to take the profile curve of the cylinder as general as we wish.

To do that, let $\alpha(s)$ be any plane curve in $\mathbb{R}^3$, which we take arc length parameterized, where $s \in \mathrm{I} \subset \mathbb{R}$, I being an interval. Let P be the plane where $\alpha(s)$ is living and let $\vec{v}$ be a unit vector orthogonal to that plane. Then the cylinder over α is the parameterized surface given by

$$\mathrm{C}_\alpha = \{\mathbf{X}(s, t) = \alpha(s) + t\,\vec{v}\ :\ s \in \mathrm{I},\ t \in \mathbb{R}\}.$$

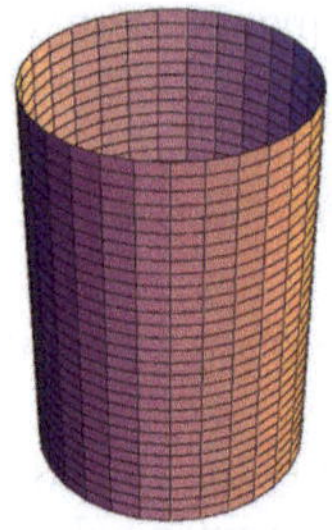

A right circular cylinder

Let us do now the following experiment. In the plane (s, t), a straight line passing through the origin with slope θ is given by $(s = au, t = bu)$, where $a = \sin\theta$ and $b = \cos\theta$. Take its image by $\mathbf{X}$ in the cylinder C_α to find the curve

$$\gamma(u) = \mathbf{X}(au, bu) = \alpha(au) + bu\,\vec{v}.$$

The experiment is to compute its Frenet trihedron $\{T_\gamma, N_\gamma, B_\gamma\}$, and its curvatures functions κ_γ and τ. To do that, we start from the Frenet frame $\{T(s), N(s)\}$ of the plane curve $\alpha(s)$, which is arc length parameterized, and its curvature function is $\kappa(s)$. The parameters of $\alpha(s)$ and the straight line are related by $ds = a\,du$. As for the arc length parameter $\bar{s}$ of the curve γ we have that

$$\frac{d\gamma}{du} = a\,\frac{d\alpha}{ds} + b\,\vec{v},$$

so that

$$\left|\frac{d\gamma}{du}\right|^2 = 1 \quad \Rightarrow \quad d\bar{s} = du = \frac{1}{\sin\theta}\,ds.$$

Then we find the unit tangent vector field

$$T_\gamma(s) = \frac{d\gamma}{du} = \sin\theta\,T(s) + \cos\theta\,\vec{v}$$

and the unit normal to γ

$$N_\gamma(s) = \frac{1}{\sin\theta}\,J((\sin\theta)\,T(s)) = N(s).$$

We differentiate again and apply the first Frenet equation of both curves to get

$$\frac{dT_\gamma}{d\bar{s}} = \frac{dT_\gamma}{ds}\,\frac{ds}{d\bar{s}} = \sin^2\theta\,\frac{dT}{ds},$$

and

$$\kappa_\gamma = \kappa \sin^2 \theta.$$

As for the binormal and torsion we have

$$B_\gamma = T_\gamma \times N_\gamma = -\cos\theta\, T(s) + \sin\theta\, \vec{v}.$$

Finally, a new differentiation jointly with the third Frenet equation of γ yields

$$\tau = \kappa \sin\theta \cos\theta.$$

Consequently, γ is a Lancret helix and $\frac{\tau}{\kappa} = \cot\theta$ is the slope of the chosen straight line.

We know that right cylinders over plane curves are flat surfaces, that is, surfaces with zero Gaussian curvature. Geometrically, they are surfaces locally isometric to a plane (the isometry is provided by the map **X**). As a consequence, the images of straight lines (which are geodesics of the plane) map to geodesics of the corresponding cylinder. Moreover, all of geodesics of a cylinder are obtained in this way. Summarizing, we have:

Any Geodesic of a Right Cylinder Over a Plane Curve Is a Lancret Helix
It seems natural to state the following question:

Can we say that any Lancret helix is a geodesic of a right cylinder over a plane curve?

The Answer Is Affirmative
Indeed, if γ is a Lancret helix then the ratio between its torsion and curvature is constant, i.e.,

$$\frac{\tau_\gamma}{\kappa_\gamma} = \cot\theta.$$

Let α be a plane curve arc length parameterized whose curvature function will be given by

$$\kappa_\alpha = \frac{1}{\sin^2\theta}\, \kappa_\gamma.$$

Let C_α be the cylinder over α, where we take the geodesic corresponding to the slope θ, that is,

$$\beta(u) = \alpha(\sin\theta\, u) + \cos\theta\, u\, \vec{v}.$$

Remember that $\vec{v}$ is a unit vector orthogonal to the plane where α is lying, and so it defines the generatrices of the cylinder. As we saw before, β has the same arc length, same curvature and same torsion as γ. From the fundamental theorem of the theory of space curves, there is a rigid motion bringing one to another.

Then we have shown the following result:
Let γ be a regular curve in $\mathbb{R}^3$. The following statements are equivalent:

1. **γ is a helix, i.e., it makes a constant angle with a fixed direction.**
2. **The ratio between its curvature and torsion is constant, the slope.**
3. **γ is a geodesic of a right cylinder built over a plane curve.**

Furthermore, we have done a procedure to build, up to rigid motions, the whole family of helices in $\mathbb{R}^3$, which depends on two parameters:

(i) A real valued function playing the role of the curvature function of the profile curve of the cylinder where the geodesic lies down.
(ii) A real number playing the role of the slope of the geodesic of the cylinder.

1.4 Examples

We will exhibit some interesting examples of curious helices appearing in different contexts (see [7] and [9]). An overall procedure to build helices consists of developing the exposed ideas, which essentially allows us to solve geometrically the natural equations of such curves. Any Lancret helix γ satisfies that $\tau/\kappa = \cot\theta$. Then assuming that the axis of γ is the z-axis, we have that $dz = \cos\theta\, ds$, where s is, as usual, its arc length parameter. It is well known that the projection of γ over the xy-plane is a curve α whose arc length parameter satisfies that $ds_\alpha = \sin\theta\, ds$ and its curvature function is given by $\kappa_\alpha = \kappa \csc^2\theta$. By solving the natural equations of α we find that

$$\varphi_\alpha = \int \kappa_\alpha\, ds_\alpha = \csc^2\theta \int \kappa\, ds,$$

φ_α standing for the angle between the tangent to α and the x-axis. Therefore, we obtain the following arc length parameterization of the Lancret helix γ

$$\begin{aligned} x(s) &= \sin\theta \textstyle\int_0^s \cos\left[\csc^2\theta \int_0^u \kappa(v)\, dv\right] du \\ y(s) &= \sin\theta \textstyle\int_0^s \sin\left[\csc^2\theta \int_0^u \kappa(v)\, dv\right] du \\ z(s) &= \cos\theta\, s. \end{aligned}$$

Elliptic helices. They are ubiquitous in Nature. For instance, the appear in protein chains, antennas and nanotubes. Such helices are geodesics, with slope $h = p/n$, of right cylinders over ellipses. Let us start from an ellipse

$$\alpha(u) = (r_1 \cos u, r_2 \sin u, 0)$$

in the $z = 0$ plane. Its arc length function is $s(u) = r_2 \int_0^u \sqrt{1 - \lambda \sin^2 \theta}\, d\theta$, appearing a second kind elliptic integral, where $\lambda = 1 - (r_1^2/r_2^2)$ and $r_2 > r_1$. Such a function, and its inverse, can be numerically treated by using some standard mathematical software. It is worth quoting the work by A. Gray [49], who have got numerical solutions for elliptic helices with Mathematica.

Spherical helices. An epicycloid is a plane curve drawn by the point of a circle moving, without slipping, on the exterior of another circle. If γ_h is a Lancret helix, with slope $h = \cot\theta$, then $\tau = h\,\kappa$. We know that γ_h is lying on a sphere of radius r if and only if $R^2 + (TR')^2 = r^2$, where $R = 1/\kappa$ and $T = 1/\tau$. We solve that equations to get

$$\kappa = \frac{1}{\sqrt{r^2 - h^2 s^2}}, \qquad \tau = \frac{h}{\sqrt{r^2 - h^2 s^2}}.$$

Combining the expression of the curvature with the algorithm to build helices, we find a parameterization of a spherical Lancret helix, which we project down over the xy-plane to obtain an epicycloid α_h with radii $r \cos\theta$ and $r \sin^2\theta/2$ (see [94] for more details).

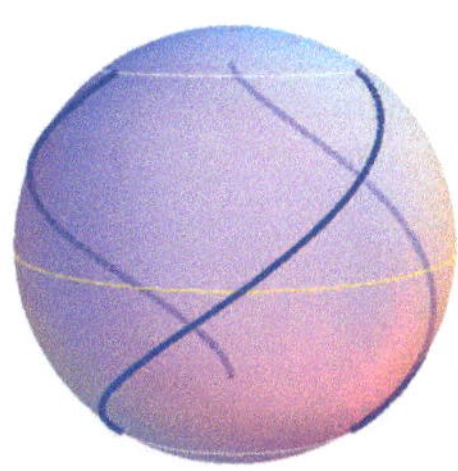

A spherical helix

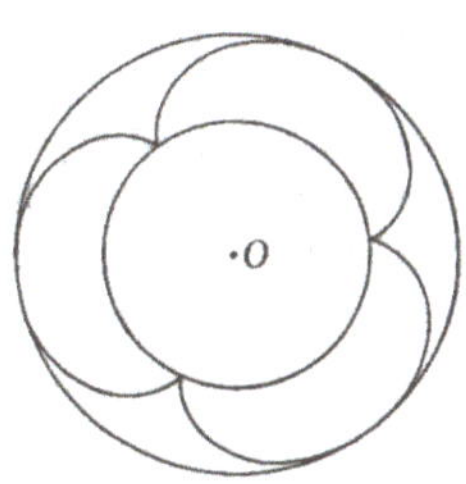

An epicycloid

Conical helices. They appear as geodesics of right cylinders whose profile curve is either a logarithmic or an Archimedean spiral. The former can be parameterized by

$$\alpha(u) = (r\, u \cos(c \ln u), r\, u \sin(c \ln u)) , \quad u > 0,$$

u standing for the arc length parameter. On the cylinder $\phi(u, v) = \alpha(u) + v\, \vec{e}_3$, consider the geodesic with slope $h = \cot\theta = p/n$ to find the curve

$$\gamma_h(t) = (r\, nt \cos(c \ln(nt)), r\, nt \sin(c \ln(nt)), p\, t) ,$$

which is a conical helix lying on a cone of equation $x^2 + y^2 = \dfrac{r^2 n^2}{p^2} z^2$. The same can be done starting from an Archimedean spiral.

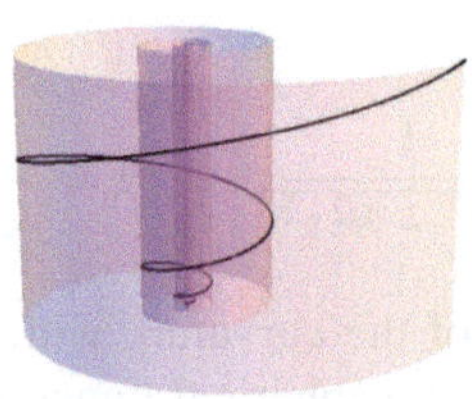

A logarithmic conical Lancret curve

An Archimedean conical Lancret curve

Lancret helices over a Cornu spiral. The transition curve is a classical problem in the construction of roads and railways, and deals with connecting straight and circular sections. The problem addressed by transition curves is the sudden change in centripetal acceleration as a moving vehicle enters a fixed radius curve. At slow speeds this is rather trivial, but as speed increases it becomes a significant jerk, unpleasant for people, stressful for hardware. A transition curve eliminates the jerk by providing a continuous change from one centripetal acceleration to another. This problem is solved by using the *Cornu spirals*, also known as clothoids or Euler spirals.

They are plane curves whose curvature function is proportional to its arc length. By writing, for instance, $\kappa(s) = s$, we can easily solve the natural equations as follows. Let

$$\theta(s) = \int_0^s \kappa(t)\,dt = \frac{1}{2}s^2 + c,$$

be the angle function making the tangent with x-axis. With a suitable choice of axes, we can assume, without of generality, that $c = 0$. Then the unit tangent to the Cornu spiral is

$$T(s) = \left(\cos\frac{s^2}{2}, \sin\frac{s^2}{2}\right),$$

which yields the following arc length parameterization of the curve

$$\alpha(s) = \left(\int_0^s \cos\frac{u^2}{2}\,du, \int_0^s \sin\frac{u^2}{2}\,du\right).$$

The integrals there are known as Fresnel integrals. On the cylinder $\phi(s, v) = \alpha(s) + v\vec{e}_3$, built over $\alpha(s)$, we take any geodesic. As it is determined by its slope $h = p/n$, we find the following parameterized Lancret helix

$$\gamma_h(t) = \left(\int_0^{nt} \cos\frac{u^2}{2}\,du, \int_0^{nt} \sin\frac{u^2}{2}\,du, pt\right),$$

which can be used to model the vine stalks.

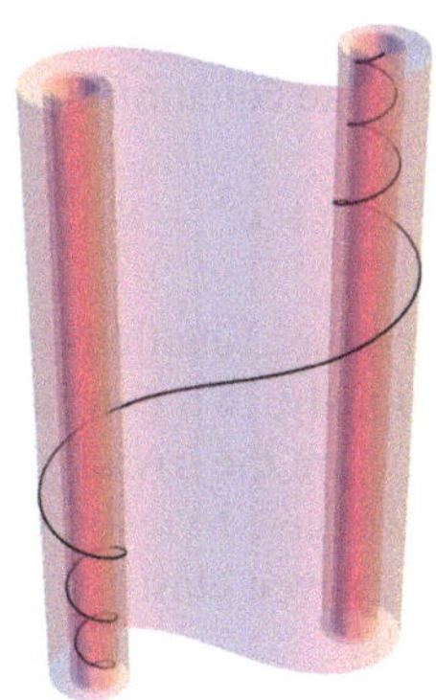

A Cornu Lancret curve

That Cornu spiral is the first of a long list of arc length parameterized clothoids by

$$\alpha(s) = \left(\int_0^s \cos\left(\frac{u^{n+1}}{n+1}\right) du, \int_0^s \sin\left(\frac{u^{n+1}}{n+1}\right) du \right), \quad n \in \mathbb{N}.$$

You can naturally construct Lancret helices as geodesics of cylinders whose profile curves are in that family of clothoids.

Lancret helices over Poleni curves. The syntractrices are classical curves studied by Giovanni Poleni circa 1729 to build a device to compute logarithms. A syntractrix is closely related to the tractrix in the sense that the former is described by the points on the tangent to the latter. Let P be a point of a tractrix $\mathcal{T}$, and let M be the corresponding point on its asymptote. We know the distance $|PM|$ is constant. We pay attention to the points, say N, on the tangent such that $|NM| = r|PM|$, r standing for a fixed number. Then when P sketches $\mathcal{T}$, the point M draws its asymptote, then N describes a syntractrix. In particular, when $r = 2$, the syntractrix is known as either *Poleni curve* or *courbe des forçats*.
Later, the Poleni curve will appear again in the *elastic problem*. This problem, stated by Jordanus de Nemore in the thirteenth century, drew the attention of a significant number of great scientists and mathematicians. It was considered by Galileo in 1638, Hooke in 1678, James Bernoulli in 1691–1694, variationally stated by Daniel Bernoulli in 1742, and finally solved by Euler in 1744. From a current viewpoint, it looks for plane curves which are stationary, i.e., critical points of the following functional

$$\mathcal{F}_\lambda(\alpha) = \int_\alpha (\kappa^2(s) + \lambda)\, ds, \qquad \lambda \in \mathbb{R}.$$

As we said before, Euler solved the problem finding the nine plane elastic forms. The corresponding Euler-Lagrange equation writes as follows

$$2\kappa'' + \kappa^3 - \lambda\kappa = 0, \tag{1.1}$$

whose solutions yield the curvature functions of the plane elastic curves. That differential equation has, obviously, two constant curvature solutions, those for straight lines and circles, which are the trivial examples of Euler elastics. As for finding non-trivial solutions, we have to resort to the elliptic functions to obtain the curvatures of the remaining seven elastic forms. Among them, the only non-periodic solution is that given by

$$\kappa(s) = 2q\ \text{sech}(qs), \quad q = \sqrt{2\lambda},$$

which is the curvature function of the Poleni curve. We know that, using quadratures, the natural equations of any plane curve can be solved. In other words, once we know the curvature function of a plane curve, by means of quadratures we explicitly obtain the curve. As for the Poleni curve we find the following parametric equations

$$u(s) = s - \frac{2}{q}\tanh(qs), \qquad v(s) = \frac{2}{q}\operatorname{sech}(qs).$$

We can check that curve is arc length parameterized, so that we can construct, in $\mathbb{R}^3$, the right cylinder over it, where we will take a geodesic with slope $h = p/n$, to find the Lancret helix parameterized by

$$\gamma_h(t) = \left(nt - \frac{2}{q}\tanh(qnt), \frac{2}{q}\operatorname{sech}(qnt), pt \right).$$

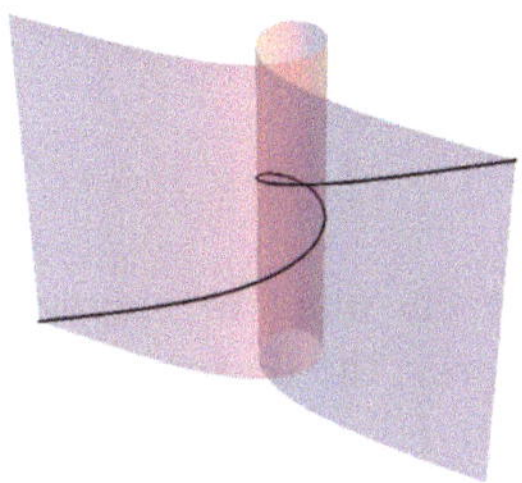

A Poleni Lancret curve

1.5 Magnetic Fields on a Riemannian Manifold

A *magnetic vector field* on an n-dimensional Riemannian manifold M, with metric $g = \langle\, , \rangle$, is a *closed* 2-*form* $F \in \Lambda^2(M)$, i.e., $dF = 0$. By using the metric, we can define the *Lorentz force* of the magnetic vector field as follows

$$\langle \phi(X), Y \rangle = F(X, Y), \qquad \phi \in TM.$$

The dynamics of an *electric charge* moving in a magnetic vector field is described by the curves $\gamma(t)$, which we will call *magnetic curves* and are solutions of the following *Lorentz force equation*

$$\nabla_{\gamma'} \gamma'(t) = \phi(\gamma'(t)), \tag{1.2}$$

as it is well known, where ∇ stands for the Riemannian connection of (M, g). Naturally, the above equation, for the trivial magnetic vector field $F = 0$, turns into the geodesics equation of (M, g), which, as is well known, are characterized as critical points of the energy functional and they represent the trajectories of a free fall particle (only influenced by the gravity field). On the other hand, the antisymmetry of the Lorentz forces ϕ (a consequence of the same property of F)

allows us to assure that, as geodesics, the magnetic curves have constant speed. In fact.

$$\frac{d}{dt}\langle\gamma',\gamma'\rangle = 2\langle\nabla_{\gamma'}\gamma',\gamma'\rangle = 2\langle\phi(\gamma'),\gamma'\rangle = 0.$$

An interesting class of magnetic vector fields is that formed by those satisfying that $\nabla F = 0$. They will be called *uniform magnetic vector fields*, which obviously have parallel Lorentz force, $\nabla\phi = 0$. Its importance lies, partly, in its connection with the *classical Landau-Hall problem*. The object of the problem is to look for the dynamics of a charge in a uniform magnetic vector field defined in a constant curvature surface. This is equivalent to integrate the Lorentz force equation (1.2). This problem has an easy solution by using simple geometric arguments. In fact, we have to observe that, in a Riemannian surface, (M^2, g), any 2-form can be written as the product of a differentiable function and the 2-form Ω_2 which gives the area element on the surface, i.e.,

$$\Lambda^2(M) = \{f\,\Omega_2 \,:\, f \in C^\infty(M)\}.$$

Now it is easy to see that the uniform magnetic vector fields correspond to constant functions. In other words, the uniform magnetic vector fields are constant multiples of the area element. If $F = c\,\Omega_2$, $c \in \mathbb{R}$, is one of them, we can compute its Lorentz force as follows. At any point $p \in M$ and for any unit vector $\vec{x} \in T_pM$, let $J\vec{x} \in T_pM$ be the vector obtained under a $\pi/2$ rotation with a orientation that, at least locally, can be chosen such that $\Omega_2(\vec{x}, J\vec{x}) = 1$. Then we have that

$$\phi(\vec{x}) = \langle\phi(\vec{x}, J\vec{x}\rangle\, J\vec{x} = F(\vec{x}, J\vec{x})\, J\vec{x} = c\,\Omega_2(\vec{x}, J\vec{x})\, J\vec{x} = c\, J\vec{x}.$$

As a consequence, $\phi = c\,J$ and it can be combined with the Lorentz force equation to find that

$$\nabla_{\gamma'}\gamma'(t) = c\,J(\gamma'(t)).$$

Finally, if we consider the first Frenet equation and bear in mind that the unit normal of the curve $\gamma(t)$ is $J(\gamma'(t))$, we obtain that

$$\nabla_{\gamma'}\gamma'(t) = c\,J(\gamma'(t)) = \kappa\,J(\gamma'(t)),$$

κ standing for the curvature function of the curve. It should be noted that the magnetic curve has been taken arc length parameterized, which is not restrictive, because we know that the magnetic curves are travelled around at constant speed.

Summarizing, *on any surface, the dynamics of a charge in a uniform magnetic vector field, with Lorentz force $\phi = c\,J$, develops trough curves with constant curvature c*. In the special case of a surface of constant curvature, we know that, up to topology, we are talking about $\mathbb{R}^2$, $\mathbb{S}^2$ or $\mathbb{H}^2$. In these spaces, the curves of constant curvature are well known, so that the classical Landau-Hall problem has been reduced to an easy geometric problem whose solution is known.

In general, we will call Landau-Hall problem to that of finding the dynamics of a charge under the action of any magnetic vector field in a Riemannian manifold. Or equivalently, finding the solutions of (1.2).

Now we invite the reader to the variational party of the magnetic curves. We wish to see the Lorentz force equation as the Euler-Lagrange associated with a certain energy function. To determine such an action, we first use that, at least locally, on a manifold, any closed form is also exact. Then, given the magnetic vector field F on the manifold M, at any point of M we can find a neighborhood U where $F = d\omega$ (in physics terms, it is said that the magnetic vector field comes from a potential ω). This potential can be globally defined on certain manifolds. For instance, when the second cohomology group is trivial, $H^2(M) = 0$, then any closed form is also exact, and therefore, in those spaces any magnetic vector field globally comes from a potential. In any case, as we just said, for any magnetic vector field, on any manifold, a potential is always locally guaranteed.

If we take two points $p, q \in U$ and consider the space of curves in U connecting them

$$\Lambda_{pq} = \{\alpha : [a, b] \to U \ : \alpha(a) = p, \ \ \alpha(b) = q\},$$

we can define an energy action $\mathcal{L} : \Lambda_{pq} \to \mathbb{R}$ as follows

$$\mathcal{L}(\alpha) = \frac{1}{2} \int_\alpha \langle \alpha', \alpha' \rangle \, dt - \int_\alpha \omega(\alpha') \, dt.$$

The tangent space to Λ_{pq} at α can be seen in M identified with the space of vector fields along the curve α which vanish at the ending points. If $Z \in T_\alpha \Lambda_{pq}$, then we can consider a curve in Λ_{pq} passing trough α and with tangent vector Z. For instance,

$$\Gamma : (-\varepsilon, \varepsilon) \to \Lambda_{pq}, \quad z \mapsto \Gamma_z,$$

where Γ_z, viewed in M, is the following curve

$$\Gamma_z(t) = \exp_{\alpha(t)} z \, Z(t).$$

It is obvious that

$$\Gamma_0(t) = \exp_{\alpha(t)} 0 = \alpha(t) \quad \Longleftrightarrow \quad \Gamma_0 = \alpha,$$

and

$$\frac{\partial}{\partial z}(\Gamma_z(t))\bigg|_{z=0} = (d\exp_{\alpha(t)})_0 \, Z(t) = Z(t) \quad \Longleftrightarrow \quad (\Gamma_z')(0) = Z.$$

Now we can compute the first variation $\delta\mathcal{L}(\alpha) : T_\alpha\Lambda_{pq} \to \mathbb{R}$ of the above energy action by using, as always, the integration by parts. Then we have that

$$\delta\mathcal{L}(\alpha)[Z] = \frac{\partial}{\partial z}\mathcal{L}(\Gamma_z)\bigg|_{z=0} = Z\left(\mathcal{L}(\Gamma_z)\right) = \int_\alpha \langle \nabla_Z \alpha', \alpha'\rangle\, dt - \int_\alpha Z(\omega(\alpha'))\, dt.$$

To do the integration by parts and use the boundary conditions $Z(a) = Z(b) = 0$, we note that

$$\begin{aligned}
\langle \nabla_Z \alpha', \alpha'\rangle = \langle \nabla_{\alpha'} Z, \alpha'\rangle &= \frac{d}{dt}\langle Z, \alpha'\rangle - \langle Z, \nabla_{\alpha'} \alpha'\rangle, \\
Z(\omega(\alpha')) = d\omega(Z, \alpha') - \alpha'(\omega(Z)) &= F(Z, \alpha') - \alpha'(\omega(Z)) \\
&= -\langle Z, \phi(\alpha')\rangle - \alpha'(\omega(Z)).
\end{aligned}$$

Therefore

$$\delta\mathcal{L}(\alpha)[Z] = \left(\langle Z, \alpha'\rangle\right)_a^b - \int_\alpha \langle Z, \nabla_{\alpha'} \alpha'\rangle\, dt + \int_\alpha \langle Z, \phi(\alpha')\rangle\, dt + \left[\alpha'(\omega(Z))\right]_a^b.$$

The boundary conditions yield

$$\delta\mathcal{L}(\alpha)[Z] = -\int_\alpha \langle Z, \nabla_{\alpha'} \alpha' - \phi(\alpha')\rangle\, dt,$$

so that a curve α is critical for the energy action $\mathcal{L}$, that is, $\delta\mathcal{L}(\alpha)[Z] = 0$, for all $Z \in T_\alpha\Lambda_{pq}$, if and only if

$$\nabla_{\alpha'} \alpha' - \phi(\alpha') = 0,$$

i.e., α is a solution of the Lorentz force equation, in other words, it is a magnetic curve.

Since we have met the equations of the Lorentz force, from which the magnetic curves are solutions, as well as the Euler-Lagrange associated with an energy action, the above argument shows the variational nature of the magnetic curves. In any case, questions arise derived from the existence of a potential. Here we have some of them:

(i) First, the existence of a potential, globally defined, for any magnetic vector field is directly related to the topology of the space where the magnetic vector field is defined. As we said before, if $H^2(M)$ is trivial, then any 2-closed form is exact and therefore any magnetic vector field comes from a potential globally defined on M.

(ii) The existence of a potential, for a certain magnetic vector field, is not unique. In fact, if F has a potential, for instance ω, then we have $F = d\omega$. But it is evident that for any function $f \in C^{\infty}(M)$, the 1-form $\omega + df$ always is a potential for F.

(iii) As a consequence, the old aspiration of showing the variational nature of the magnetic curves from a principle which involves neither the existence of potentials nor the topology of the space remains to be an aspiration. Nevertheless, this problem has been solved in some special cases in low dimension (see [12] and [14]).

On the other hand, there exists another problem naturally related to the Lorentz force equation (1.2). It is clear that the existence and uniqueness of the magnetic curves, as well as for geodesics, is guaranteed from a point and a tangent vector at that point. More precisely, for each $p \in M$ and $v \in T_pM$, there exists a unique inextensible magnetic curve passing trough p with velocity vector v. Then one can fall into the temptation of stating the following question: *Given a non-trivial magnetic vector field, F, on a Riemannian manifold M, is there an affine connection on M whose geodesics are the magnetic curves of F?*

In some sense, we are proposing to see whether the dynamics of charges in a magnetic vector field can be interpreted, in the same space, as free falling particles in another gravitational field. Naturally, the answer to that problem is negative, as was shown in [12]: *If F is a non-trivial magnetic vector field on (M, g), then there exists no affine connection on M whose geodesics are the magnetic curves of (M, g, F).*

1.6 Killing Fields

This section is taken from [69]. Let $\gamma(t)$ be a non-null immersed curve in a 3-dimensional Lorentzian space form M with sectional curvature c and let $v(t) = |\gamma'(t)|$ be the speed of γ. Let us consider a variation of γ, $\Gamma = \Gamma(t, z) : I \times (-\varepsilon, \varepsilon) \to M$ with $\Gamma(t, 0) = \gamma(t)$. In particular one can choose $\varepsilon > 0$ in such a way that all t-curves of the variation have the same causal character as that of γ. Associated with Γ there are two vector fields along Γ, $V(t, z) = \frac{\partial \Gamma}{\partial z}(t, z)$ and $W(t, z) = \frac{\partial \Gamma}{\partial t}(t, z)$. In particular $V(t) = V(t, 0)$ is the variational vector field along γ and $W(t, z)$ is the tangent vector fields of the t-curves. We will use the notation $V = V(t, z)$, $v = v(t, z)$, $\kappa = \kappa(t, z)$, etc. with the obvious meanings. Also, if s denotes the arc length parameter of the t-curves, we will write $v(s, z)$, $V(s, z)$, $\kappa(s, z)$, etc. for the corresponding reparameterizations.

A straightforward but long computation allows us to obtain formulas for $\dfrac{\partial v}{\partial z}(t, 0)$, $\dfrac{\partial \kappa^2}{\partial z}(t, 0)$ and $\dfrac{\partial \tau^2}{\partial z}(t, 0)$ which we collect, along with another standard identity, in the following lemma.

Lemma 1.6.1 *With the above notation, the following assertions hold:*

(1) $[V, W] = 0$;

(2) $\frac{\partial v}{\partial z}(t, 0) = -\varepsilon_1 g v$, *with* $g = \langle \bar{\nabla}_T V, T \rangle$;

(3) $\frac{\partial \kappa^2}{\partial z}(t, 0) = 2\varepsilon_2 \langle \bar{\nabla}_T^2 V, \bar{\nabla}_T T \rangle + 4\varepsilon_1 g \kappa^2 + 2\varepsilon_2 \langle R(V, T)T, \bar{\nabla}_T T \rangle$;

(4) $\frac{\partial \tau^2}{\partial z}(t, 0) = -2\varepsilon_2 \langle \frac{1}{\kappa} \bar{\nabla}_T^3 V - \frac{\kappa'}{\kappa^2} \bar{\nabla}_T^2 V + \varepsilon_1(\varepsilon_2 \kappa + \frac{c}{\kappa}) \bar{\nabla}_T V - \varepsilon_1 \frac{c\kappa'}{\kappa^2} V, \tau B \rangle$,

where $\langle , \rangle$ *denotes the Lorentzian metric of* M *and* $\kappa' = \frac{\partial \kappa}{\partial t}(t, 0)$.

Without loss of generality we can assume γ to be arclength parameterized. A vector field $V(s)$ along γ, which infinitesimally preserves unit speed parameterization (that means $\frac{\partial v}{\partial z}(t, 0) = 0$ for a V-variation of γ) is said to be a *Killing vector field along* γ if this evolves in the direction of V whithout changing shape, only position. In other words, the curvature and torsion functions of γ remain unchanged as the curve evolves. Hence Killing vector fields along γ are characterized by the equations

$$\frac{\partial v}{\partial z}(t, 0) = \frac{\partial \kappa^2}{\partial z}(t, 0) = \frac{\partial \tau^2}{\partial z}(t, 0) = 0, \tag{1.3}$$

and this is well defined in the sense that it does not depend on the V-variation of γ one chooses to compute the derivatives involved in Eq. (1.3). In fact, we use Lemma 1.6.1 and (1.3) to see that V is a Killing vector field along γ if and only if it satisfies the following conditions:

$$\begin{aligned} &\text{(a) } \langle \bar{\nabla}_T V, T \rangle = 0, \\ &\text{(b) } \langle \bar{\nabla}_T^2 V, N \rangle + \varepsilon_1 c \langle V, N \rangle = 0, \\ &\text{(c) } \langle \frac{1}{\kappa} \bar{\nabla}_T^3 V - \frac{\kappa'}{\kappa^2} \bar{\nabla}_T^2 V + \varepsilon_1(\varepsilon_2 \kappa + \frac{c}{\kappa}) \bar{\nabla}_T V - \varepsilon_1 c \frac{\kappa'}{\kappa^2} V, \tau B \rangle = 0. \end{aligned} \tag{1.4}$$

In particular, the solutions of (1.4) form a 6-dimensional linear space.

Now when M is simply connected, since the restriction to γ of any Killing field $\tilde{V}$ of M is a Killing vector field along γ, one concludes from a well known dimension argument, the following lemma.

Lemma 1.6.2 *Let* M *be a complete, simply connected, Lorentzian space form and* γ *a non-null immersed curve in* M. *A vector field* V *on* γ *is a Killing vector field along* γ *if and only if it extends to a Killing field* $\tilde{V}$ *on* M.

1.7 Magnetic Nature of Lancret Helices

Remember that a *magnetic vector field* on an n-dimensional Riemannian manifold M, with metric $g = \langle \, , \rangle$, is a *closed 2-form* $F \in \Lambda^2(M)$, i.e., $dF = 0$. This concept has special interest when $\dim(M) = 3$, where we are able to visualize the magnetic vector fields in their most classical sense, i.e., as a genuine divergence zero vector fields. To do that, we use the well known musical isomorphisms, which work on any Riemannian manifold, but when $\dim(M) = 3$ they provide a natural isomorphism between fields and 1-forms as follows:

$$\flat : \mathfrak{X}(M) \to \Lambda^1(M), \quad (X^\flat)(Y) = \langle X, Y \rangle, \quad \forall X, Y \in (X)(M),$$

and its inverse

$$\sharp : \Lambda^1(M) \to \mathfrak{X}(M), \quad \langle \omega^\sharp, Y \rangle = \omega(Y), \quad \text{for any } \omega \in \Lambda^1(M) \text{ and } \quad Y \in (X)(M).$$

We also have the Hodge star operator, which in $\dim(M) = 3$ gives an isomorphism between 2-forms and 1-forms

$$\star : \Lambda^2(M) \longleftrightarrow \Lambda^1(M).$$

In both cases they are defined by

$$\eta \wedge \star\beta = \langle \eta, \beta \rangle \, \Omega_3,$$

Ω_3 standing for the volume form of (M, g).

Then we can define a new isomorphism between 2-forms and vector fields given by

$$\Lambda^2(M) \to \mathfrak{X}(M), \quad F \mapsto (\star F)^\sharp .$$

Is is easy to show that the inverse is written down as

$$\mathfrak{X}(M) \to \Lambda^2(M), \quad V \mapsto \star\left(V^\flat\right).$$

The last 2-form, using the contraction operator, can be written as follows

$$\star\left(V^\flat\right) = \iota_V \Omega_3.$$

To show it, let $\{E_1, E_2, E_3\}$ be a local basis of orthonormal vector fields an let $\{\omega_1, \omega_2, \omega_3\}$ its dual basis. Then we have

$$V = f_1 \, E_1 + f_2 \, E_2 + f_3 \, E_3, \quad V^\flat = f_1 \, \omega_1 + f_2 \, \omega_2 + f_3 \, \omega_3,$$

so that

$$\star\left(V^\flat\right) = f_1\,\omega_2 \wedge \omega_3 - f_2\,\omega_1 \wedge \omega_3 + f_3\,\omega_1 \wedge \omega_2.$$

On the other hand, the components in that basis of the 2-form $\iota_V\Omega_3$ are given by

$$(\iota_V\Omega_3)\,(E_1, E_2) = \Omega_3(V, E_1, E_2) = f_3\,\Omega_3(E_3, E_1, E_2) = f_3,$$

$$(\iota_V\Omega_3)\,(E_1, E_3) = \Omega_3(V, E_1, E_3) = f_2\,\Omega_3(E_2, E_1, E_3) = -f_2,$$

$$(\iota_V\Omega_3)\,(E_2, E_3) = \Omega_3(V, E_2, E_3) = f_1\,\Omega_3(E_1, E_2, E_3) = f_1.$$

Since in a 3-dimensional Riemannian manifold we are identifying 2-forms and vector fields, it will be very useful to understand the translation of closed 2-forms in order to get the magnetic vector field in a classical sense, i.e., as a very vector field. To do that we will compute the differential of $\mathrm{i}_V\Omega_3$ to check whether it is closed. Then

$$d\,(\iota_V\Omega_3) = d\,(f_1\,\omega_2 \wedge \omega_3 - f_2\,\omega_1 \wedge \omega_3 + f_3\,\omega_1 \wedge \omega_2)\,,$$

and so

$$d\,(\iota_V\Omega_3) = df_1 \wedge \omega_2 \wedge \omega_3 - df_2 \wedge \omega_1 \wedge \omega_3 + df_3 \wedge \omega_1 \wedge \omega_2.$$

Therefore we have got

$$d\,(\iota_V\Omega_3) = (E_1(f_1) + E_2(f_2) + E_3(f_3))\;\omega_1 \wedge \omega_2 \wedge \omega_3 = \operatorname{div}(V)\,\Omega_3.$$

We have shown that, from the above isomorphism, closed 2-forms are in one-to-one correspondence with divergence zero vector fields, in other words, **in dimension three, a magnetic vector field is a divergence zero vector field. In particular, any Killing vector field is a magnetic vector field**.

1.7.1 The Lorentz Force in Dimension Three

The *Lorentz force* ϕ of the magnetic vector field F is the map $\phi : TM \to TM$ defined by

$$\langle\phi(X), Y\rangle = F(X, Y).$$

As we have just seen, when $\dim(M) = 3$ a magnetic vector field is either a closed 2-form F or a divergence zero vector field V. Then, to compute the Lorentz force of V we will use the volume form to define the vector product by

$$\langle X \times Y, Z\rangle = \Omega_3(X, Y, Z), \qquad X, Y, Z \in (X)(M).$$

Now the Lorentz force of the magnetic vector field V is given by

$$\langle \phi(X), Y\rangle = F(X, Y) = (\iota_V \Omega_3)(X, Y) = \Omega_3(V, X, Y) = \langle V \times X, Y\rangle,$$

from which we deduce that

$$\boxed{\phi(X) = V \times X} \tag{1.5}$$

1.7.2 Circular Helices as Magnetic Curves

In the Euclidean 3-space $\mathbb{R}^3$, equipped with the usual coordinates (x, y, z), we consider the vector field $V = \partial_z$, which is obviously a constant length Killing field, also called *infinitesimal translation*. It provides a magnetic vector field in $\mathbb{R}^3$ and we look for the corresponding magnetic curves.

In physics terms, this problem can be stated as follows: *Compute the trajectories in $\mathbb{R}^3$ of a charge in the magnetic vector field $V = \partial_z$.*

To solve it, we take a circular helix on a cylinder

$$\gamma_{ab}(t) = (a \cos t, a \sin t, bt), \qquad a, b \in \mathbb{R},$$

whose axis is $V = \partial_z$. Now we do the following elementary computations:

$$\gamma'_{ab}(t) = (-a \sin t, a \cos t, b),$$
$$\gamma''_{ab}(t) = (-a \cos t, -a \sin t, 0) = \partial_z \times \gamma'_{ab}(t).$$

Then we have shown that any circular helix lying on a right circular cylinder with axis ∂_z is a trajectory of the magnetic vector field $V = \partial_z$.

Now we will see that the converse also holds. Assume that the curve $\gamma(t) = (x(t), y(t), z(t))$ is a trajectory of the magnetic vector field $V = \partial_z$. Then it should be a solution of the Lorentz force equation $\gamma''(t) = \partial_z \times \gamma'(t)$, which implies that

$$(x''(t), y''(t), z''(t)) = (-y'(t), x'(t), 0).$$

From the third equation we find that $z(t) = nt + p$ and, without loss of generality, we can take $p = 0$. The remaining equations yield

$$x'(t) = -y(t), \qquad y'(t) = x(t),$$

so that the plane curve $(x(t), y(t))$ is a circle. As a consequence, the starting curve is a geodesic lying on a right circular cylinder and therefore it is a circular helix.

1.7.3 Lancret Helices as Magnetic Curves

It seems now natural to state the following questions:

1. *What happen with non-circular Lancret helices?*
2. *Are there magnetic vector fields whose magnetic trajectories are Lancret helices?*
3. *Is any Lancret helix a magnetic curve of a magnetic vector field?*

To answer these questions, let $\gamma(t) = (x(t), y(t), z(t))$ be a Lancret helix. We can assume, making a change of coordinates if necessary, that its axis is ∂_z. Furthermore, $\gamma(t)$ can be taken with constant speed $\|\gamma'(t)\| = r$, otherwise it could not be a magnetic curve. As we know, $\gamma(t)$ should be geodesic in a right cylinder over a curve $\alpha(t)$ lying in the plane $z = 0$. Let φ be the slope of $\gamma(t)$, i.e., the constant angle of $\gamma'(t)$ with its axis and write $a = r\cos\varphi$. Then the third coordinate of the curve is nothing but $z(t) = at+b$. It is well known that the curvature functions of both curves are related by

$$\kappa_\gamma = \kappa_\alpha \sin^2\varphi.$$

The speed and acceleration of the plane curve are given by

$$\|\alpha'(t)\| = r\sin\varphi, \qquad \alpha''(t) = (r^2\sin^2\varphi)\,\kappa_\alpha(t)\,\frac{J(\alpha'(t)}{r\sin\varphi},$$

J being a direct $\pi/2$-rotation. Writing down the second equation in coordinates we find that

$$x''(t) = -(r\sin\varphi)\kappa_\alpha(t)\,y'(t), \quad y''(t) = (r\sin\varphi)\kappa_\alpha(t)\,x'(t).$$

Both equations, along with $z(t) = at + b$, show that

$$\gamma''(t) = [(r\sin\varphi)\kappa_\alpha(t)]\,\partial_z \times \gamma'(t).$$

As a consequence, *the Lancret helix is a magnetic curve of a magnetic vector field of the form* $V = f(x, y)\,\partial_z$, where the function f, called the potential, along the Lancret helix, is given by $f(\gamma(t)) = (r\sin\varphi)\kappa_\alpha(t)$.

It should be understood that any vector field in $\mathbb{R}^3$ of the form $V(x, y, z) = f(x, y)\,\partial_z$ has zero divergence and therefore it is a magnetic vector field. This fact will be used to compute the magnetic curves. The Lorentz force equation for that magnetic field is given by

$$\gamma''(t) = f\,(x(t), y(t))\;\partial_z \times \gamma'(t),$$

which can be rewritten as $\gamma(t) = (x(t), y(t), z(t))$. Then we obtain

$$x''(t) = -f(x(t), y(t))\,y'(t), \quad y''(t) = f(x(t), y(t))\,x'(t), \quad z''(t) = 0. \tag{1.6}$$

The third equation yields $z(t) = at + b$, for any pair of constants $a, b \in \mathbb{R}$. Furthermore, taking the projection of the curve on the plane $z = 0$, that is, $\alpha(t) = (x(t), y(t), 0)$, the right cylinder over $\alpha(t)$ is

$$\mathrm{C}_\alpha = \{\mathrm{X}(t, v) = \alpha(t) + v\,\partial_z\ :\ t, v \in \mathbb{R}\}.$$

Then the original curve can be seen as a geodesic of this cylinder. Indeed, we see that $\gamma(t) = \alpha(t) + (at + b)\partial_z = \mathrm{X}(t, at + b)$. Actually, we have shown that $\gamma(t)$ is a Lancret helix. Summing up, the magnetic curves of the magnetic vector fields of the form $V(x, y, z) = f(x, y)\,\partial_z$ are Lancret helices in $\mathbb{R}^3$.

In a physical context, we have shown that **the trajectories of a charge in $\mathbb{R}^3$ in the magnetic vector field** $V(x, y, z) = f(x, y)\,\partial_z$ **are Lancret helices.**

Now we wish to relate the essential elements (curvature function of the base curve and slope) of a Lancret helix with the potential function $f(x, y)$ of the magnetic vector field. To do that we remind that $\|\gamma'(t)\| = r$, so that the speed of the projection is given by $\|\alpha'(t)\| = \sqrt{r^2 - a^2} = r \sin\varphi$, where φ is the slope of the Lancret helix. Moreover, we use the two first equations of (1.6) to write the acceleration vector field of the projection as

$$\alpha''(t) = f(x(t), y(t))\,J(\alpha'(t)),$$

J standing for the $\pi/2$ positive rotation. Then we obtain the curvature function of the projected curve

$$\kappa_\alpha(t) = \frac{f(x(t), y(t))}{r \sin\varphi}.$$

Summarizing, *the dynamics of a charge in $\mathbb{R}^3$ moving in a magnetic vector field of the form $V(x, y, z) = f(x, y)\,\partial_z$ is described by trajectories which are Lancret helices of slope φ, lying on cylinders over plane curves $\alpha(t)$, with curvature function $\kappa_\alpha(t)$ and crossed at speed r, in such a way that* $(r \sin\varphi)\,\kappa_\alpha(t) = f(x(t), y(t))$.

1.7.4 The Magnetic Algorithm and Its Applications

We are going to provide a mechanism to build the trajectories of a charge and beautiful examples of magnetic flows associated with Lancret helices.

First step. Start with a differentiable function $h : \mathbb{R}^2 \to \mathbb{R}$ defined on the plane $z = 0$;

Second step. Remove its critical points to get a restricted function $h_{|U} = h : U \subset \mathbb{R}^2 \to \mathbb{R}$, where $U = \{p \in \mathbb{R}^2 : \nabla h(p) \neq 0\}$;

Third step. Use the Implicit Function Theorem to find a foliation in $U \subset \mathbb{R}^2$ by level curves of the given function. In other words, for each $p \in U$ we have the level curve $\mathcal{C}_p = \{q \in U \ : h(q) = h(p)\}$ and then $U = \bigcup_{p \in U} \mathcal{C}_p$;

Fourth step. Choose an initial slope φ and a travel speed r to define the potential of the magnetic vector field, $f : U \subset \mathbb{R}^2 \to \mathbb{R}$, by

$$f(p) = (r \sin\varphi)\, \kappa_{\mathcal{C}_p}(p);$$

and

Last step. The magnetic curves of the field $V(x, y, z) = f(x, y)\, \partial_z$ are the Lancret helices with slope φ lying on cylinders built over the curves $\mathcal{C}_p$ crossed at speed r. Then, any charge under that field moves along one of those Lancret helices.

We are able to compute the curvature function of any curve of the above foliation in terms of h. To do that, we know that the gradient of the function, $\nabla h = (h_x, h_y)$, is normal to the curves of the foliation, so that a foliation corresponds to the flow of the field $X(x, y) = (h_y, -h_x)$. Then the curvature of those curves is given by

$$\kappa(x, y) = \frac{1}{|\nabla h|^3} \langle \nabla_X X, \nabla h \rangle = \begin{pmatrix} h_y & -h_x \end{pmatrix} \begin{pmatrix} h_{xx} & h_{xy} \\ h_{xy} & h_{yy} \end{pmatrix} \begin{pmatrix} h_y \\ -h_x \end{pmatrix},$$

or briefly,

$$\kappa(x, y) = \frac{\mathrm{Hes}(h)(X, X)}{|\nabla h|^3}.$$

Some Examples That algorithm allows us to exhibit nice examples of charges crossing along Lancret helices under the influence of those magnetic vector fields. Some of them are listed below.

(1) Let us start with the function $h : \mathbb{R}^2 \to \mathbb{R}$, defined by $h(x, y) = x^2 - y^2$. The origin is obviously the only critical point, which can be removed, although it is not essential. The level curves, where the asymptotes $x = \pm y$ can be included (even the origin), are hyperbolas. These can be seen as integral curves of the vector field $X(x, y) = (-2y, -2x)$. It is clear that $\mathrm{Hes}(X, X) = 8y^2 - 8x^2$,

and so we can define the potential

$$f(x, y) = (r \sin\varphi) \frac{8y^2 - 8x^2}{(4x^2 + 4y^2)^{3/2}}.$$

Then the magnetic curves of the vector field $V(x, y, z) = f(x, y)\,\partial_z$, are Lancret helices crossed at speed r, with slope φ on the cylinders built over the above foliation of hyperbolas.

(2) Remove a point of the plane, for instance the origin, to define the function $h : \mathbb{R}^2 - \{(0, 0)\} \to \mathbb{R}$ given by

$$h(x, y) = \frac{\sqrt{x^2 + y^2}}{\arctan(y/x)}.$$

Observe that its level curves are Archimedean spirals. Now consider the vector field $V(x, y, z) = f(x, y)\,\partial_z$, where the potential is given by

$$f(x, y) = (r \sin\varphi) \frac{\mathrm{Hes}(h)(X, X)}{|\nabla h|^3}.$$

We conclude that a charge in that field is moving, at constant speed r, along a Lancret helix with slope φ, on a cylinder built over a foliation by those spirals.

(3) We remove now a straight line of the plane, for instance, a coordinate axis, to define the function $h : \mathbb{R}^2 - \{(0, y) : y \in \mathbb{R}\} \to \mathbb{R}$ by

$$h(x, y) = \frac{(x^2 + y^2)^2 + y^2}{x^2}.$$

Its level curves are Booth lemniscatas (also known as hyperbolic lemniscatas). This flow of curves is generated by the vector field $X(x, y) = (h_y, -h_x)$, which is influenced by the magnetic vector field $V(x, y, z) = f(x, y)\,\partial_z$, where

$$f(x, y) = (r \sin\varphi) \frac{\mathrm{Hes}(h)(X, X)}{|\nabla h|^3}.$$

Its magnetic curves are Lancret helices of constant speed r, with slope φ, lying on cylinders over the above lemniscatas.

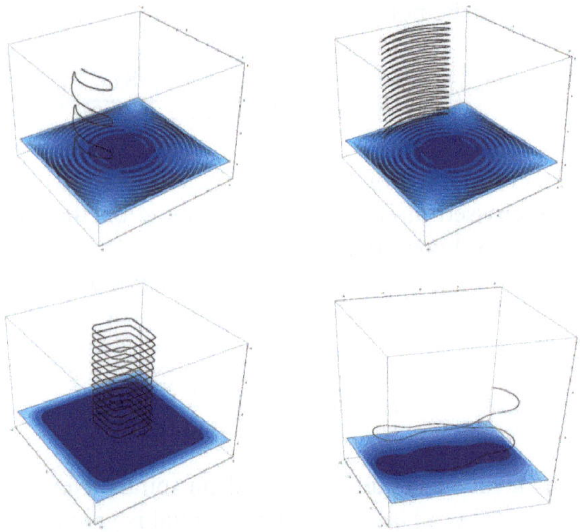

Some examples of Booth lemniscatas

1.8 Lancret Helices as Solution of a Variational Problem

In this section we look for Lancret helices which are solutions of variational problems associated with functionals whose densities are concerned with the geometrical invariants of the dynamical variables: speed, curvature and torsion. In other words, we search for Lancret helices as critical points of certain energies defined on a family of curves (see [7] and [9]).

Let us consider the space of curves in $\mathbb{R}^3$ connecting two fixed points $p, q \in \mathbb{R}^3$, that is,

$$\Lambda = \{\gamma : [a, b] \to \mathbb{R}^3 \; : \; \gamma(a) = p, \; \gamma(b) = q\}.$$

Now, on Λ we take the tri-parametric family of functionals

$$\mathcal{F}_{mnp} : \Lambda \to \mathbb{R}, \qquad \mathcal{F}_{mnp}(\gamma) = \int_\gamma (m + n\kappa + p\,\tau)\, ds, \quad m, n, p \in \mathbb{R}.$$

They are linear functionals in the three invariants of a space curve: speed, curvature and torsion, so that they measure the weights of length, bending and twisting in the action. We are interested in the problem of characterizing the critical points of that kind of Lagrangians. In other words, we wish to describe the dynamics governed for that family of functionals.

The Space of Curves Our first aim is to identify the tangent space to Λ at a point $\gamma \in \Lambda$. To do that, take a curve in Λ passing through γ, that is, $\Gamma : (-\varepsilon, \varepsilon) \to \Lambda$, such that $\Gamma(v) \in \Lambda$, $\forall v \in (-\varepsilon, \varepsilon)$ and $\Gamma(0) = \gamma$. The differentiability of this curve can be seen in $\mathbb{R}^3$. Indeed, from $\Gamma : (-\varepsilon, \varepsilon) \to \Lambda$, we can build a map, also written as Γ, as follows

$$\Gamma : [a, b] \times (-\varepsilon, \varepsilon) \to \mathbb{R}^3, \qquad \Gamma(t, v) = \Gamma(v)(t).$$

We will say that the curve is differentiable provided the corresponding application is differentiable in $\mathbb{R}^3$. Then we can define two vector fields along that map

$$X(t, v) = \frac{\partial \Gamma}{\partial t}(t, v), \qquad V(t, v) = \frac{\partial \Gamma}{\partial v}(t, v),$$

that we will call *longitudinal* and *transversal*, respectively. As for the tangent space at γ we have that

$$\Gamma'(0) = \frac{\partial \Gamma}{\partial v}(t, 0) = V(t, 0).$$

Furthermore, as the curves in Λ are connecting two points of M, we have that $V(a, 0) = V(b, 0) = 0$. We have found that any tangent vector to Λ at γ is viewed in $\mathbb{R}^3$ as a vector field $V(t) = V(t, 0)$, along $\gamma(t)$, which vanishes at the ending points.

The converse also holds. Given a vector field $V(t)$ along $\gamma \in \Lambda$, vanishing at the ending points, $V(a) = V(b) = 0$, we can define a curve in Λ, passing through γ with velocity vector $V(t)$. Indeed, let us write

$$\Gamma : (-\varepsilon, \varepsilon) \to \Lambda, \qquad \Gamma(v)(t) = \exp_{\gamma(t)}(v\, V(t)) = \gamma(t) + v\, V(t).$$

It is clear that

$$\Gamma(0) = \gamma, \qquad \Gamma'(0) = V.$$

As a consequence, *the tangent space to Λ at a point γ can be seen in $\mathbb{R}^3$ as the space of vector fields along $\gamma(t)$ vanishing at the ending points, i.e.,*

$$T_\gamma \Lambda = \{V(t) \in T_{\gamma(t)} M \, : \, V(a) = V(b) = 0\}.$$

The first variation formula. In a second step, we look for a useful formula for the first variation

$$\delta \mathcal{F}_{mnp}(\gamma) : T_\gamma \Lambda \to \mathbb{R}.$$

To do that, we have to know the transversal variations of the speed, curvature and torsion, because they are used to define the Lagrangian density. As for the longitudinal $X(t, v)$ and transversal $V(t, v)$ vector fields we have that

$$[X, V] = 0.$$

On the other hand, by considering the speed function $x(t, v) = \|X(t, v)\|$ of the longitudinal curves, we have the unit field

$$T(t, v) = \frac{X(t, v)}{x(t, v)}, \quad \text{or} \quad X(t, v) = x(t, v)\, T(t, v).$$

Bringing this to the above bracket, we find that $\nabla_{x\,T}\, V = \nabla_V\, (x\, T)$, which implies that

$$[V, T] = -\frac{V(x)}{x}\, T. \tag{1.7}$$

Then we obtain the following formula for the transversal variation of the speed of the longitudinal curves

$$V(x) = \frac{\partial x}{\partial v} = \frac{\partial}{\partial v}\sqrt{\langle X, X\rangle} = \frac{\langle \nabla_V\, X, X\rangle}{x} = \frac{\langle \nabla_X\, V, X\rangle}{x} = \langle \nabla_T\, V, T\rangle\, x. \tag{1.8}$$

As for the transversal variation of the curvature function of longitudinal curves, we start by computing

$$V(\kappa^2) = \frac{\partial \kappa^2}{\partial v} = V\langle \nabla_T\, T, \nabla_T\, T\rangle = 2\langle \nabla_V \nabla_T\, T, \nabla_T\, T\rangle.$$

Combining now the formula for the curvature with (1.7) we deduce that

$$\nabla_V \nabla_T\, T = \nabla_T \nabla_V\, T + \nabla_{[V,T]}\, T = \nabla_T \nabla_V\, T - \frac{V(x)}{x}\, T,$$

and so

$$\begin{aligned} V(\kappa^2) &= 2\langle \nabla_T(\nabla_T\, V + [V, T]) - \frac{V(x)}{x}\, \nabla_T\, T; \nabla_T\, T\rangle \\ &= 2\langle \nabla_T^2\, V - T\left(\frac{V(x)}{x}\right)\, T - 2\frac{V(x)}{x}\, \nabla_T\, T; \nabla_T\, T\rangle. \end{aligned}$$

Then we have got

$$2\kappa\, V(\kappa) = 2\kappa\langle \nabla_T^2\, V, N\rangle - 4\frac{V(x)}{x}\kappa^2,$$

which yields the required formula for the variation of the curvature

$$V(\kappa) = \langle \nabla_T^2\, V, N\rangle - 2\langle \nabla_T\, V, T\rangle\, \kappa. \tag{1.9}$$

To find the variation of the torsion we have to work a bit more. Let us start with the transversal variation of the first Frenet equation to find that

$$\kappa \nabla_V N = \nabla_V \nabla_T T - V(\kappa)\, N = \nabla_T \nabla_V T + \nabla_{[V,T]} T - V(\kappa)\, N,$$

which combines with (1.7) to obtain

$$\nabla_V N = \frac{1}{\kappa} \langle \nabla_T^2 V, B \rangle\, B - \langle \nabla_T V, N \rangle\, T. \tag{1.10}$$

Now we compute the transversal variation of the second Frenet equation

$$\begin{aligned} \nabla_V \nabla_T N &= \nabla_T \nabla_V N + \nabla_{[V,T]} N \\ &= -V(\kappa)\, T - \kappa \nabla_V T + V(\tau)\, B + \tau \nabla_V B. \end{aligned}$$

Computing the binormal components in the above expression we obtain

$$V(\tau) = \langle \nabla_T \nabla_V N + \nabla_{[V,T]} N, B \rangle + \kappa \langle \nabla_V T, B \rangle.$$

Finally, bringing here (1.7) and (1.10) we find the following useful formula

$$V(\tau) = -T \left(\frac{1}{\kappa} \langle \nabla_T^2 V, B \rangle \right) - \langle \nabla_T V, \tau T + \kappa B \rangle. \tag{1.11}$$

The Euler-Lagrange equations. From the formulas (1.8), (1.9) and (1.11), and using repeatedly integration by parts, we deduce the first variation formula for the functionals $\mathcal{F}_{mnp}$

$$\begin{aligned} \delta \mathcal{F}_{mnp}(\gamma)[V] &= \frac{\partial}{\partial v} \int_{\Gamma(t,v)} (m + \kappa(t,v) + \tau(t,v))\, ds \Big|_{v=0} \\ &= \frac{\partial}{\partial v} \int_a^b (m + \kappa(t,v) + \tau(t,v))\, x(t,v)\, dt \Big|_{v=0} \\ &= n \int_a^b \frac{\partial \kappa}{\partial v} x\, dt + p \int_a^b \frac{\partial \tau}{\partial v} x\, dt + \int_a^b (m + n\kappa + p\tau) \frac{\partial x}{\partial v}\, dt \\ &= n \int_\gamma \frac{\partial \kappa}{\partial v} ds + p \int_\gamma \frac{\partial \tau}{\partial v} ds + \int_a^b (m + n\kappa + p\tau) \langle \nabla_T V, T \rangle\, x\, dt \\ &= n \int_\gamma \frac{\partial \kappa}{\partial v} ds + p \int_\gamma \frac{\partial \tau}{\partial v} ds + \int_\gamma (m + n\kappa + p\tau) \langle \nabla_T V, T \rangle\, ds. \end{aligned}$$

We will compute separately the three terms of the right hand side.

First Term

$$\int_\gamma \frac{\partial\kappa}{\partial v}\,ds = \int_\alpha \langle \nabla_T \nabla_T\, V, N\rangle\, ds - 2\int_\alpha \langle \nabla_T\, V, T\rangle\, \kappa\, ds$$
$$= \int_\gamma \frac{d}{ds}\langle \nabla_T\, V, N\rangle\, ds - \int_\gamma \langle \nabla_T\, V, \nabla_T\, N\rangle\, ds - 2\int_\gamma \frac{d}{ds}\left(\langle V, T\rangle\, \kappa\right) ds$$
$$+ 2\int_\gamma \langle V, \nabla_T\, T\rangle \kappa\, ds + 2\int_\gamma \kappa' \langle V, T\rangle ds.$$

By integrating by parts in the second term of the right hand side of the last expression we obtain

$$\int_\gamma \frac{\partial\kappa}{\partial v}\,ds = \int_\gamma \langle V, \nabla_T^2\, N\rangle ds + 2\int_\gamma \langle V, \nabla_T\, T\rangle \kappa ds + 2\int_\gamma \kappa' \langle V, T\rangle ds$$
$$+ \int_\gamma \frac{d}{ds}\left(\langle \nabla_T\, V, N\rangle - \langle V, \nabla_T\, N\rangle - 2\langle V, T\rangle\, \kappa\right) ds.$$

We write the above formula as follows

$$\int_\gamma \frac{\partial\kappa}{\partial v}\,ds = \int_\gamma \langle \Omega_1(\gamma), V\rangle ds \;+\; \text{an exact integral}, \tag{1.12}$$

where

$$\Omega_1(\gamma) = \nabla_T^2\, N + 2\kappa \nabla_T\, T + 2\kappa'\, T.$$

Second Term From (1.11) we see that the first term is an exact integral, so that

$$\int_\gamma \frac{\partial\tau}{\partial v}\,ds = \int_\gamma \langle V, \nabla_T(\tau T + \kappa B)\rangle - \int_\gamma \frac{d}{ds}\left(\frac{1}{\kappa}\langle \nabla_T^2\, V, B\rangle + \langle V, \tau T + \kappa B\rangle\right) ds.$$

As above, this can be written as

$$\int_\gamma \frac{\partial\tau}{\partial v}\,ds = \int_\gamma \langle \Omega_2(\gamma), V\rangle ds \;+\; \text{an exact integral}, \tag{1.13}$$

where

$$\Omega_2(\gamma) = \nabla_T(\tau T + \kappa B).$$

Third Term A simple computation yields

$$\begin{aligned}\int_\gamma (m+n\kappa+p\tau)\langle\nabla_T V, T\rangle\, ds &= \int_\gamma \frac{d}{ds}\left((m+n\kappa+p\tau)\langle V, T\rangle\right) ds \\ &\quad - \int_\gamma (n\kappa'+p\tau')\langle V, T\rangle\, ds \qquad (1.14)\\ &\quad - \int_\gamma (m+n\kappa+p\tau)\langle V, \nabla_T T\rangle\, ds.\end{aligned}$$

Rewrite as

$$\int_\gamma (m+n\kappa+p\tau)\langle\nabla_T V, T\rangle\, ds = \int_\gamma \langle\Omega_3(\gamma), V\rangle ds \;+\; \text{an exact integral}, \qquad (1.15)$$

where

$$\Omega_3(\gamma) = -(n\kappa'+p\tau')\,T - (m+n\kappa+p\tau)\,\nabla_T T.$$

The Formula Gathering together the above computations we have

$$\delta\mathcal{F}_{mnp}(\gamma)[V] = \int_\gamma \langle\Omega(\gamma), V\rangle ds + \mathcal{B}(\gamma, V)|_a^b,$$

where Ω is the Euler-Lagrange operator given by

$$\begin{aligned}\Omega(\gamma) &= n\,\Omega_1(\gamma) + p\,\Omega_2(\gamma) + \Omega_3(\gamma)\\ &= n\left(\nabla_T^2 N + 2\kappa\nabla_T T + 2\kappa' T\right) + p\,(\nabla_T(\tau T + \kappa B)) \qquad (1.16)\\ &\quad - (n\kappa'+p\tau')\,T - (m+n\kappa+p\tau)\,\nabla_T T.\end{aligned}$$

Now we use the Frenet equations

$$\begin{aligned}\nabla_T T &= \kappa\, N,\\ \nabla_T N &= -\kappa\, T - \tau\, B,\\ \nabla_T B &= \tau\, N,\end{aligned}$$

to write the EulerLagrange operator as

$$\Omega(\gamma) = (-n\,\tau^2 + p\,\kappa\tau - m\,\kappa)N + (-n\,\tau' + p\,\kappa')B. \qquad (1.17)$$

On the other hand, we collect all of exact integrals in a boundary term given by

$$\mathcal{B}(\gamma, V) = -\frac{p}{\kappa}\langle\nabla_T^2 V, B\rangle + n\langle\nabla_T V, N\rangle + (n\,\tau - p\,\kappa)\langle V, B\rangle + m\langle V, T\rangle. \tag{1.18}$$

Henceforth we are going to find suitable boundary conditions to characterize the critical points of the functionals. To do that it will be enough to consider second order conditions, in other words, we will work on the space of clamped curves $\Lambda_c \subset \Lambda$, which is made of curves connecting two fixed points and having the same unit tangent and normal vectors there. We can also say that Λ_c is the family of curves that have the same Frenet trihedron at the points they are connecting. Then we restrict the functionals to Λ_c and will continue denoting by

$$\mathcal{F}_{mnp} : \Lambda_c \longrightarrow \mathbb{R}.$$

It is clear that the boundary conditions yield

$$\mathcal{B}(\gamma, V)|_a^b = 0.$$

Then the first variation formula is

$$\delta\mathcal{F}_{mnp}(\gamma)[V] = \int_\gamma \langle\Omega(\gamma), V\rangle ds, \tag{1.19}$$

where the Euler-Lagrange operator Ω was given by (1.17).

Now we are going to characterize the critical points of the functionals $\mathcal{F}_{mnp}$ on the space Λ_c, that is, the curves $\gamma \in \Lambda_c$ such that $\delta\mathcal{F}_{mnp}(\gamma) = 0$. It is clear that $\Omega(\gamma) = 0$ implies that $\delta\mathcal{F}_{mnp}(\gamma) = 0$ and therefore the curve is a critical point. The converse also holds. Indeed, if $\gamma \in \Lambda_c$ is critical point, then $\delta\mathcal{F}_{mnp}(\gamma)[V] = 0$, and now we take a vector field of the form $V = f(s)\,\Omega(\gamma)$, where f is a positive function such that f, f' y f'' vanish at a and en b. Then we obtain

$$\delta\mathcal{F}_{mnp}(\gamma)[V] = \int_\gamma f(s)\|\Omega(\gamma)\|^2\, ds = 0,$$

so that $\Omega(\gamma) = 0$.

Summarizing, the critical points of $\mathcal{F}_{mnp}$ on Λ_c are just the solutions of $\Omega(\gamma) = 0$, in other words, the curves in $\mathbb{R}^3$ whose geometrical invariants, curvature and torsion, are solutions of the following equations

$$\begin{aligned} -n\,\tau^2 + p\,\kappa\tau - m\,\kappa &= 0, \\ -n\,\tau' + p\,\kappa' &= 0. \end{aligned} \tag{1.20}$$

They can be completely integrated and their solutions, according to the values of the parameters, are gathered in the following table:

m	n	p	Trajectories	Conditions
$\neq 0$	$= 0$	$= 0$	Geodesics	$\kappa = 0$
$= 0$	$= 0$	$\neq 0$	Circles	$\kappa = \text{cte}, \quad \tau = 0$
$= 0$	$\neq 0$	$= 0$	Plane curves	$\tau = 0$
$\neq 0$	$\neq 0$	$= 0$	Circular helices	$\kappa = -\frac{n}{m}\tau^2$
$\neq 0$	$= 0$	$\neq 0$	Circular helices	any $\kappa, \quad \tau = \frac{m}{p}$
$= 0$	$\neq 0$	$\neq 0$	Lancret helices	$\tau = \frac{p}{n}\,\kappa$
$\neq 0$	$\neq 0$	$\neq 0$	Circular helices	$\kappa = -\frac{na^2}{m+ap}, \ \tau = \frac{ma}{m+ap}, \ a \in \mathbb{R} - \{-\frac{m}{p}\}$

Some interesting consequences can be obtained, two of them deserve our comments.

(I) Closed Solutions The variational problem for closed curves is a especial case within clamped ones. It is obvious that a Lancret helix (in particular, a circular helix), which is not contained in a plane, can not be closed, because the coordinate in the axis direction is a linear function, and so it is not periodic. Then, from the above table, the only variational models admitting closed solutions are those concerning with

- The total torsion, or (*twisting*), $m = n = 0$ and $p = 1$, $\mathcal{T}(\gamma) = \int_\gamma \tau\, ds$, whose critical points are circles.
- The total curvature, or (*bending*), $m = p = 0$ and $n = 1$, $\mathcal{B}(\gamma) = \int_\gamma \kappa\, ds$, whose critical points are plane curves.

The latter is of special interest as it can be compared to the situation described in the classical well known Fenchel theorem (see [40]):

If γ is a closed curve in $\mathbb{R}^3$, then its total curvature satisfies that

$$\mathcal{B}(\gamma) = \int_\gamma \kappa\, ds \geq 2\pi.$$

Furthermore, **equality holds if and only if γ is a plane convex curve**. In other words, the minima of the *bending* function for closed curves is achieved on plane convex curves.

In our case we see that, not only the minima, but also **the critical points of the *bending* functional are plane curves**. Moreover, **the critical values of the *bending* functional satisfy the following Dirac quantization principle: they come as integer multiples of a fixed value**.

For a better understanding of the above result we have to remind that, when considering curves in $\mathbb{R}^3$, the curvature function is signed, i.e., $\kappa \geq 0$. In particular, if we take a critical point γ of the total curvature function, we know that it should be a plane curve which is seen as a space curve, and therefore with non-negative curvature. To know the corresponding critical value we have to compute

$$\mathcal{B}(\gamma) = \int_\gamma |\kappa_\gamma|\, ds,$$

κ_γ standing for the curvature function of γ seen as a plane curve. Now, starting from γ, we build a new plane and convex curve, γ_c, geometrically obtained from γ by a symmetrization procedure. That can be done by reflecting its concave pieces with respect to the tangent lines at the inflection points. In other words, γ_c is an arc length parameterized curve which is determined, up to rigid motions in the plane, by its curvature function defined by

$$\kappa_{\gamma_c}(s) = |\kappa_\gamma(s)|.$$

As a consequence, the critical value is given by

$$\mathcal{B}(\gamma) = \int_\gamma |\kappa_\gamma|\, ds = 2\pi\, \mathrm{i}(\gamma_c),$$

where $\mathrm{i}(\gamma_c) \in \mathbb{N}$ is the rotation index of the convex curve γ_c. We conclude that the set of critical values for the *bending* functional, on the closed space curves, is just the discrete set $\{2\pi k \,:\, k \in \mathbb{N}\}$.

(II) Lancret Solutions The most interesting case is concerned with a functional where $m = 0$. We know that the trajectories are *Lancret helices on cylinders built over any plane curves whose slope is determined by the ratio between the weights of the* twisting *and the* bending. In this case, we will see that the energy of a Lancret helix should be given by a natural multiple of a fixed energy value. Consequently, we can say that the critical values of the energy $\mathcal{F}_{0np} \equiv \mathcal{F}_{np}$ are quantized *à la Dirac*. Indeed, bearing in mind the above considerations and the formulas relating the curvature and the torsion of a Lancret helix, γ, being κ_α the curvature of the corresponding transversal section, we obtain

$$\mathcal{F}_{np}(\gamma) = n \int_\gamma \kappa_\gamma\, ds + p \int_\gamma \tau\, ds = \sqrt{n^2 + p^2} \int_\alpha |\kappa_\alpha| ds.$$

Then, for an energy $\mathcal{F}_{np}$, its critical values are given, up to a constant which depends on the weights of the energy, by the absolute total curvature of the corresponding transversal section. By considering, for instance, closed plane curves, and for each one, α, the arc length parameterized convex curve α_c, which is determined by its curvature function $\kappa_{\alpha_c} = |\kappa_\alpha|$, then we find the critical value of the energy defined on any Lancret helix with slope p/n

$$\mathcal{F}_{np}(\gamma) = 2\pi\sqrt{n^2 + p^2}\,\mathrm{i}(\alpha_c),$$

$\mathrm{i}(\alpha_c) \in \mathbb{N}$ standing for the rotation index of α_c.

1.9 Uniqueness of the Energy Action

In [42] the authors proposed a variational model to describe protein chains which is governed by an energy action whose density does not depend on the torsion, but is linear in the curvature. According to previous discussion, that model lies in the above table when $m\,n \neq 0$ and $p = 0$. The authors motivate that choice based on the following arguments:

(1) On the one hand, the helicoidal structure of the proteins involves the choice of an energy action whose extremals are helices.
(2) On the other hand, the main result of [42], which is stated as follows: *If the extremals of $\mathcal{F}(\gamma) = \int_\gamma F(\kappa(s))\,ds$ are circular helices, then $F(\kappa) = m + n\kappa$, with $m\,n \neq 0$.*

These reasons show some inconsistency with the concept of helix. Indeed, the authors first pointed out the helicoidal nature of protein chains. Secondly, they requested that extremals are circular helices. However, the helicoidal structures we usually find in Nature are far from being circular helices. They refused to consider that the space of critical points contains forms as natural as elliptic, spherical or conical helices.

Furthermore, there are strong arguments for not exclude, a priori, the torsion function in the density energy when one wishes to build a variational model with helicoidal extremals. Let us mention some of them:

(i) Firstly, we know that length, curvature and torsion, as geometrical invariants, are able to characterize, up to rigid motions, a space curve which is viewed as the central line of a protein chain. Then there is no reason to exclude the torsion from the energy density.
(ii) Secondly, We have just found a model (see [7]) based on a linear combination of bending and twisting whose space of extremals is made up of Lancret helices whose slope is determined from the weights of both total curvatures. That model is more accurate than that proposed in [42]. Indeed, the space of solutions of the former includes configurations such that Lancret elliptical, spherical and conical helices. The model $n\,p \neq 0$ and $m = 0$ describes that kind of solutions, as well as others which are essential in understanding not only the conformation of protein chains, but also in other contexts such as antennas or nanotechnology.

(iii) Finally, it should be pointed out that the torsion, in particular the total torsion, is a main ingredient in Calugareanu-White's equation (see [28] and [106]), which play a crucial role in the DNA theory.

Then it seems natural to state the following problem: *Determine the energy densities $F(\kappa(s), \tau(s))$ such that the energy action $\mathcal{F}(\gamma) = \int_\gamma F(\kappa(s), \tau(s))\, ds$ admits the Lancret helices, with a suitable slope, as extremals.*

We have just known a solution of that problem. Indeed, the energy action associated with the density $F(\kappa, \tau) = n\kappa + p\tau$ has a space of extremals that of the Lancret helices with slope $\omega = p/n$. In [7] we conjectured the uniqueness of that solution. More precisely, if $\mathbf{L}_\omega$ is the moduli space of Lancret helices with slope ω, then we state the following conjecture: *the space of extremals of $\mathcal{F}(\gamma) = \int_\gamma F(\kappa(s), \tau(s))\, ds$ is $\mathbf{L}_\omega$ if and only if $F(\kappa, \tau) = n\kappa + p\tau$, with $\omega = p/n$.*

That conjecture has been affirmatively solved in [9]. The solution arises when computing the Euler-Lagrange equations, under suitable boundary conditions, of the functional

$$\mathcal{F}(\gamma) = \int_\gamma F(\kappa(s), \tau(s))\, ds.$$

To do that, we follow the procedure just used for linear operators. Let us consider the space Λ of curves connecting two points and $\mathcal{F} : \Lambda \to \mathbb{R}$. The computation of $\delta\mathcal{F}(\gamma) : T_\gamma\Lambda \to \mathbb{R}$, $\gamma \in \Lambda$, is made by means of integration by parts, involving the Eqs. (1.8), (1.9) and (1.11). Then we obtain

$$\delta\mathcal{F}(\gamma)[V] = \int_\gamma \langle \Omega(\gamma), V\rangle\, ds + \mathcal{B}(\gamma, V)|_{\partial\gamma},$$

where the Euler-Lagrange operator, Ω, is of the form

$$\Omega(\gamma) = \nabla_T P, \quad P = (F - 2\kappa F_\kappa - \tau F_\tau)T + \kappa F_\tau B - \nabla_T(F_\kappa N) + \nabla_T\left(\frac{F_\tau'}{\kappa} B\right).$$

As for the term coming from the exact integrals we have got that

$$\mathcal{B}(\gamma, V) = \langle \nabla_T^2 V, \frac{F_\tau}{\kappa} B\rangle + \langle \nabla_T V, F_\kappa N - \frac{F_\tau'}{\kappa} B\rangle + \langle V, P\rangle,$$

where $F_\kappa = \partial F/\partial\kappa$, $F_\tau = \partial F/\partial\tau$ and the prime means derivative with respect to the arc length of the curve γ.

To vanish the boundary terms, we choose convenient boundary conditions. Indeed, it is enough taking curves connecting two fixed points, so that they have the same Frenet frames at the ending points. Then we can characterize the extremals γ of $\mathcal{F}(\gamma) = \int_\gamma F(\kappa(s), \tau(s))\, ds$ as the solutions of $\Omega(\gamma) = 0$. Finally, we use the Frenet equations to compute the components of $\Omega(\gamma)$. The components of this operator are in the unit normal and binormal directions, which we vanish to find the asked Euler-Lagrange equations

$$-\kappa F + (\kappa^2 - \tau^2)F_\kappa + +2\kappa\tau F_\tau + F''_\kappa + \left(\frac{\tau}{\kappa}F'_\tau\right)' + \tau\left(\frac{F'_\tau}{\kappa}\right)' = 0, \tag{1.21}$$

$$\tau F'_\kappa + \frac{\tau^2}{\kappa}F'_\tau + (\tau F_\kappa - \kappa F_\tau)' - \left(\frac{F'_\tau}{\kappa}\right)'' = 0. \tag{1.22}$$

By assuming that the functional $\mathcal{F}(\gamma) = \int_\gamma F(\kappa(s), \tau(s))\,ds$ admits as an extremal a Lancret helix with slope ω, we use (1.22) to obtain

$$\omega(\kappa F'_\kappa + \tau F'_\tau) + (\kappa F_\kappa + \tau F_\tau)' - \left(\frac{F'_\tau}{\kappa}\right)'' = 0.$$

It is clear that

$$\kappa F'_\kappa + \tau F'_\tau = (\kappa F_\kappa + \tau F_\tau)' - (\kappa' F_\kappa + \tau' F_\tau) = (\kappa F_\kappa + \tau F_\tau)' - F',$$

which yields

$$\left(\frac{F'_\tau}{\kappa}\right)' = -\omega F + (\omega\kappa + \tau)F_\kappa + (\omega\tau - \kappa)F_\tau + K,$$

K being a constant. This is combined with (1.21) to get

$$(1+\omega^2)\kappa F = (1+\omega^2)\kappa^2 F_\kappa + (1+\omega^2)\kappa\tau F_\tau + \omega K\kappa + F''_\kappa + \omega F''_\tau. \tag{1.23}$$

Remark The Eq. (1.23) is satisfied on any extremal which is a Lancret helix with slope ω. By writing $H(\kappa) = F(\kappa, \omega\kappa)$, we see that $H_\kappa = F_\kappa + \omega F_\tau$ and therefore $H''_\kappa = F''_\kappa + \omega F''_\tau$. Now take a Lancret helix with slope ω being a solution of the differential equation $H''_\kappa = 0$. We translate it into another in the curvature function of the transversal section of the cylinder where the Lancret helix is a geodesic. We realize that circular helices with slope ω are also solutions of that differential equation. As a consequence, if the functional $\mathcal{F}(\gamma) = \int_\gamma F(\kappa(s), \tau(s))\,ds$ admits a Lancret helix, with slope ω, as an extremal, being also solution of $H''_\kappa = 0$ (which is obvious provided the space of extremals contains $\mathbf{L}_\omega$), then the energy density should be of the form

$$F(\kappa, \tau) = \kappa F_\kappa + \tau F_\tau + A_1, \qquad A_1 = \frac{\omega K}{1+\omega^2}. \tag{1.24}$$

Furthermore

$$F_\kappa + \omega F_\tau = A_2, \qquad A_2 \text{ being a constant.} \tag{1.25}$$

Comparing (1.24) with (1.21) we obtain that

$$-\omega^2 \kappa F_\kappa + \tau F_\tau = A_1.$$

Solving this equation along with (1.25), we have

$$\kappa F_\kappa = \frac{A_2}{1+\omega^2}\kappa - \frac{A-1}{1+\omega^2}, \qquad \tau F_\tau = \frac{A_2\omega}{1+\omega^2}\tau + \frac{A_1}{1+\omega^2},$$

which shows that

$$F(\kappa, \tau) = \frac{A_2}{1+\omega^2}\kappa + \frac{A_2\omega}{1+\omega^2}\tau + A_1.$$

For a better understanding of the uniqueness result of the energy action, let us give an energy action $\mathcal{F}(\gamma) = \int_\gamma F(\kappa(s), \tau(s))\, ds$, and a real number ω. Take now the space $\mathbf{S}_\omega \subset \mathbf{L}_\omega$ of the Lancret helices, with slope ω, which are solutions of the equation $H''_\kappa = 0$, where $H(\kappa) = F(\kappa, \omega\kappa)$. Obviously, $\mathbf{S}_\omega$ contains all of circular helices with slope ω. Now, let $\mathbf{E}_F$ be the space of extremals of $\mathcal{F}(\gamma) = \int_\gamma F(\kappa(s), \tau(s))\, ds$. Then we have

(a) *If there exists $\gamma \in \mathbf{S}_\omega$ with $\gamma \in \mathbf{E}_F$, then the energy density is of the form*

$$F(\kappa, \tau) = m + n\kappa + p\tau, \qquad \omega = \frac{p}{n}.$$

(b) *If $\mathbf{L}_\omega \subset \mathbf{E}_F$, then*

$$F(\kappa, \tau) = n\kappa + p\tau, \qquad \omega = \frac{p}{n}.$$

Furthermore, in this case $\mathbf{L}_\omega = \mathbf{E}_F$.

1.10 Closed Lancret Helices

As was mentioned before, the coordinate function of any Lancret helix in the direction of its axis is linear, so it is not a periodic function. Therefore, there are no closed Lancret helices in $\mathbb{R}^3$. However, in Nature we can find many examples of *closed helicoidal structures*. Maybe that the most typical example is that of a curve wounded around a circular revolution torus. These curves are geometrically described by using the idea of *loxodrome*. Most people have heard of loxodromes, or *rhumb lines*, of a sphere. They are the solution of the well known problem: *what are the curves in a sphere making a constant angle with parallels; or equivalently, with meridians*. They are the loxodromes of a sphere and they form a double spiral around the poles (Poinsot spiral). As the stereographic projection from de north pole on the equator plane is a conformal map, the image of a loxodrome makes the same constant angle with the images of the meridians. But these are straight lines starting from the origin, then a loxodrome protects down in a logarithmic spiral, which shows that they are not closed.

The same question can be stated in a circular revolution torus. Actually, take the circle in the plane $y = 0$, centered at $(R, 0, 0)$ and radius r, where $R > r$. Now rotate it around the z-axis to obtain the following revolution surface

$$X(u, v) = ((R + r\cos u)\cos v, (R + r\cos u)\sin v, r\sin u),$$

whose meridians, $v =$ constant, are the different positions of the given curve when is rotated around the z-axis. The parallels are the circles described by the points of the given curve. As above, a new loxodromy problem arises: *compute the curves in the torus making a constant angle, φ_o, with the parallels*. It is not difficult to see that those curves are the solutions of the following differential equation

$$\frac{dv}{du} = \frac{r\cot\varphi_o}{R + r\cos u}.$$

The solutions are

$$v = v_o + \frac{2r\cot\varphi_o}{\sqrt{R^2 - r^2}} \arctan\left(\sqrt{\frac{R-r}{R+r}} \tan\frac{u}{2}\right),$$

and these curves are closed when

$$m = \frac{\sqrt{R^2 - r^2}}{r\cot\varphi_o} \in \mathbb{Q}.$$

In particular, when $m = 1$, which is equivalent to $\sin\varphi_o = r/R$, obtaining the so called *Villarceau circles*, which will play an important role later.

Forgetting the surface, either sphere or torus, the loxodromes, seen as space curves, make a constant angle with a certain flow, i.e., a certain vector field. Then consider a unit vector $\vec{v} \in \mathbb{R}^3$ which by parallel transport in $\mathbb{R}^3$ defines a vector field on the whole space with constant length. This field, which will be also denoted by $\vec{v}$, is a Killing field in $\mathbb{R}^3$ with constant length. Furthermore, its integral curves generate a flow of straight lines parallel to $\vec{v}$. In this context, we can reformulate the idea of Lancret helix. In fact, *the Lancret helices with axis $\vec{v}$ are the space curves making a constant angle with the flow of parallel straight lines generated by $\vec{v}$.*

Naturally, we can think of curves making a constant angle with the flow generated by any vector field. Then given a vector field X in $\mathbb{R}^3$, we can consider its flow, i.e., the set of integral curves, and say that a helix with axis X is a curve making a constant angle with the flow of X. Certainly, both the interest of the problem and the beauty of its solutions will closely depend on the geometric nature of the one-parameter group of transformations generated by X. It is clear that the most rigid geometric nature is that of vector fields generating one-parameter groups of isometries, that is, that of *Killing vector fields*. The simplest Killing fields in the Euclidean 3-space are those generating one-parameter groups of translations, which are precisely the above mentioned and called infinitesimal translations. Then we can say that *Lancret helices are those whose axis is an infinitesimal translation.*

Besides infinitesimal translations, the simplest Killing vector fields are the so called rotational. A *rotational Killing vector field* is that generating a one-parameter group of rotations around its axis. Without of generality, and by an appropriate choice of coordinates, we can assume that the axis is the z-axis. Then we have the one-parameter group $G = \{\phi_t : \mathbb{R}^3 \to \mathbb{R}^3 : t \in \mathbb{R}\}$, where

$$\phi_t(x, y, z) = \begin{pmatrix} x & y & z \end{pmatrix} \begin{pmatrix} \cos t & \sin t & 0 \\ -\sin t & \cos t & 0 \\ 0 & 0 & 1 \end{pmatrix}.$$

It generates the rotational Killing vector field

$$V(x, y, z) = (-y, x, 0) = -y\,\partial_x + x\,\partial_y,$$

which, by using cylindrical coordinates $(x, y, z) = (\rho \cos\theta, \rho \sin\theta, z)$, can be written as $V = \partial_\theta$.

Then we can state the following problem: *determine the helices whose axis is the rotational Killing vector field* $V = \partial_\theta$.

It should be noted that two ideas have arisen about the axis. First, the rotational Killing vector field has an axis, which we can assume to be the z-axis. On the other hand, the field itself works as the axis of the curves which make a constant angle, i.e., the helices with axis $V = \partial_\theta$.

It is clear that in a circular revolution torus with axis the z-axis its parallels are integral curves of the above rotational Killing vector field V. As a consequence, the loxodromes of such a torus are helices whose axis is a rotational Killing vector field. But this property not only holds on a circular revolution torus. In fact, if we rotate around the z-axis any curve γ in the open half-plane $P = \{(x, 0, z), x > 0\}$, we obtain a revolution surface, S_γ, whose parallels are integral curves of the field V. In other words, the parallels of S_γ belong to the flow generated by the rotational Killing vector field V. Therefore, *the loxodromes of any revolution surface are helices whose axis is a rotational Killing vector field with the same axis as the surface*.

The stated problem reduces to the following one: *Besides the loxodromes of a revolution surface, are there any other helices whose axis is a rotational Killing vector field*?

We will see that the answer is negative, providing the following result:

A curve in the Euclidean 3-space is a helix with axis a rotational Killing vector field if and only if it is a loxodrome of a revolution surface having the same axis of the helix.

To prove that, we have to see that any helix with axis a rotational Killing vector field should be a loxodrome in a revolution surface. The next procedure works, at least, locally, but another one, simple and standard, allows us to guarantee its global character. By taking cylindrical coordinates in $\mathbb{R}^3 - \{z - \text{axis}\}$, given an arc length parameterized curve $\gamma(s)$, we have

$$\gamma(s) = (\rho(s)\cos\theta(s), \rho(s)\sin\theta(s), z(s)),$$

so that

$$\gamma'(s) = \big(\rho'(s)\cos\theta(s) - \rho(s)\theta'(s)\sin\theta(s),\ \rho'(s)\sin\theta(s) \\ +\rho(s)\theta'(s)\cos\theta(s),\ z'(s)\big).$$

On the other hand, the rotational Killing vector field V along the curve is written as

$$V(s) = (-\rho(s)\sin\theta(s), \rho(s)\cos\theta(s), 0).$$

Then the angle between the curve and the vector field satisfies that

$$\cos\varphi(s) = \frac{\langle\gamma'(s), V(s)\rangle}{|\gamma'(s)|\,|V(s)|} = \frac{\rho^2(s)\,\theta'(s)}{\rho(s)} = \rho(s)\,\theta'(s),$$

because $|\gamma'(s)|^2 = (\rho'(s))^2 + (\rho(s)\,\theta'(s))^2 + z'(s)^2 = 1$. Therefore, the curve $\gamma(s)$ makes a constant angle φ_o with the rotational Killing vector field V if and only if $\rho(s)\,\theta'(s) = \cos\varphi_o$ is constant.

On the other hand, if R_ω stands for the rotation of angle ω around the z-axis, it is clear that

$$\gamma(s) = R_{\theta(s)}(\beta(s)), \qquad \beta(s) = (\rho(s), 0, z(s)).$$

Moreover, the tangent vector to the curve $\beta(s)$, in the half-plane P, satisfies that

$$|\beta'(s)|^2 = (\rho'(s))^2 + z'(s)^2 = 1 - (\rho(s)\,\theta'(s))^2 = 1 - \cos^2\varphi_o = \sin^2\varphi_o,$$

which shows that, up to $\gamma(s)$ is an integral curve of V, the curve $\beta(s)$ is regular. By rotating $\beta(s)$ around the z-axis we obtain a revolution surface, S_β, where the given curve $\gamma(s)$ is contained as a loxodrome.

Chapter 2
Lancret Helices in $\mathbb{S}^3$ and $\mathbb{H}^3$

In Chap. 1 a detailed study of Lancret helices in $\mathbb{R}^3$ was made. As such curves do not close, it was plain that we have to resort to new ambient spaces to give mathematical support to the different models of closed Lancret helices we find in Nature, such as protein chains.

Given two one variable functions, we look for a curve having them as its curvature functions. This problem, classically known as *solving the natural equations*, is worked out by means of a Riccati equation [94]. However, as for Lancret helices, those equations can be integrated by quadratures by using the fact that Lancret helices are geodesics of a cylinder (see [94]).

On the other hand, the curvature and torsion of an arc length parameterized curve are periodic functions of the arc length. This is a necessary but not sufficient condition, as a Lancret helix shows. N. V. Efimov [38] and W. Fenchel [41] stated "the closed curve problem": *Find, explicitly, necessary and sufficient conditions to guarantee that, given two periodic functions with the same period, the integral curve is closed.* This is an old and still open problem in "elementary" differential geometry.

Let us consider a Killing field along a curve in a 3-dimensional Riemannian manifold $(M, \langle , \rangle)$ of constant curvature c. This concept was first introduced by J. Langer and D. A. Singer when studying elastic curves (see [69] and [70]) and now we will use it to define Lancret helices in $(M, \langle , \rangle)$. That will yield a natural extension to $(M, \langle , \rangle)$ of Lancret helices just defined in $\mathbb{R}^3$ and allows us to prove an overall version of the Lancret theorem in $(M, \langle , \rangle)$. If the original version can be considered rather elemental, we will see that its extension to $(M, \langle , \rangle)$ reveals a subtle difference between the spherical and hyperbolic cases. The former looks like the Euclidean one, whereas the latter only presents trivial Lancret helices. The Lancret helices in a sphere will appear by using the Hopf map $\mathbb{S}^3 \to \mathbb{S}^2$, since it turns out that Lancret helices of $\mathbb{S}^3$ are geodesics of Hopf cylinders in $\mathbb{S}^3$.

The Lancret helices of $\mathbb{S}^3$ behave quite differently from the Euclidean ones. When studying the problem of the closed curve in $\mathbb{S}^3$, we will see that there exists a

A. Ferrandez Izquierdo, *Helices*, RSME Springer Series 16,
https://doi.org/10.1007/978-3-032-09663-0_2

rational one-parametric family of closed Lancret helices living in Hopf tori built on closed immersed curves in $\mathbb{S}^2$.

2.1 The Clifford Parallelism

In the above chapter we saw that Lancret helices in $\mathbb{R}^3$ can not be closed. Geometrically speaking, this is because its axis, viewed as an infinitesimal translation, does not admit closed integral curves. In other words, the Lancret helices of the Euclidean space are loxodromes on right cylinders built on plane curves. We have also seen that changing the nature of the axis, that is, taking a rotational Killing field as axis, we find closed Lancret helices, which geometrically could be viewed as loxodromes of revolution surfaces whose axis is orthogonal to the Killing field. However, whereas Lancret helices with axis a Killing translational can be obtained as solutions of a variational problem, those with axis a rotational Killing vector field, so far we do not know.

The Lancret helices in $\mathbb{R}^3$ cross a flow of parallel straightlines at a constant angle. This flow arises obviously as orbits associated to the action of a one-parametric group of translations

$$\{T_t(p) = p + t\vec{v} \; : \; t \in \mathbb{R}\},$$

where $\vec{v}$ is a vector in $\mathbb{R}^3$ which naturally defines a constant length Killing field, or infinitesimal translation. Given two straightlines R_1 and R_2 in $\mathbb{R}^3$, let us define the function

$$f_{12} : R_1 \to \mathbb{R}$$

by

$$f_{12}(p_1) = d(p_1, R_2),$$

being the distance from p_1 to R_2. Then, both straightlines are Euclidean parallel if and only if f_{12} is a constant function. Clearly, *being parallel* is an equivalence relation on the set of straight lines in $\mathbb{R}^3$. Furthermore, two of them lie in the same plane in $\mathbb{R}^3$.

The parallelism is not an exclusive property of the Euclidean space. In 1873, the English mathematician W. K. Clifford introduced a parallelism in the 3-sphere $\mathbb{S}^3$, which nowadays is known as Clifford parallelism. For convenience, we are thinking in the unit sphere $\mathbb{S}^3$, immersed in $\mathbb{R}^4$, which is identified with $\mathbb{C}^2$ as

$$\mathbb{S}^3 = \{z = (z_1, z_2) \in \mathbb{C}^2 \; : \; |z_1|^2 + |z_2|^2 = 1\}.$$

Now one can define the so called *Clifford translations* in the 3-sphere as follows: for each $t \in \mathbb{R}$ set

$$\varphi_t : \mathbb{S}^3 \to \mathbb{S}^3, \qquad \varphi_t(z_1, z_2) = (e^{it} z_1, e^{it} z_2).$$

Note that $\{\varphi_t : t \in \mathbb{R}\}$ is a one-parametric group of isometries on $\mathbb{S}^3$, which is identified with $\mathbb{S}^1$. Then we have the following action

$$\mathbb{S}^1 \times \mathbb{S}^3 \to \mathbb{S}^3, \qquad \varphi(e^{it}, z) = \varphi_t(z) = (e^{it} z_1, e^{it} z_2). \tag{2.1}$$

The orbit passing through a point $z \in \mathbb{S}^3$ is the curve $\alpha_z(t) = \varphi_t(z) = (e^{it} z_1, e^{it} z_2)$, whose tangent vector at z is iz. Therefore, the infinitesimal generator of that group is the Killing field

$$V(z) = iz = (iz_1, iz_2).$$

Observe that it is of unit length, so that it is an infinitesimal translation.

The orbits of the above action, i.e., the curves of that flow, are geodesics on $\mathbb{S}^3$. Indeed, $\alpha_z''(t) = -\alpha_z(t)$ is normal to $\mathbb{S}^3$ at any point.

Let us remember that the distance in $\mathbb{S}^3$ is defined by $d(z, w) = \arccos\langle z, w\rangle$. In other words, $d(z, w)$ is the length of the shortest path between z and w measured over the great circle of $\mathbb{S}^3$ passing through z and w. Also, for any great circle $\mathbf{C}_1$ of $\mathbb{S}^3$, define $\mathbf{d}(z, \mathbf{C}_1) = \inf\{d(z, w),\ w \in \mathbf{C}_1\}$ and for two great circles $\mathbf{C}_1$ and $\mathbf{C}_2$, $\mathbf{d}(\mathbf{C}_1, \mathbf{C}_2) = \inf\{\mathbf{d}(z, \mathbf{C}_2),\ z \in \mathbf{C}_1\}$. Observe that those orbits, great circles of $\mathbb{S}^3$, behave as the straight lines in $\mathbb{R}^3$ in the sense of the following equidistance property. By writing $\mathbf{C}_1 \equiv \{e^{it} z\}$ and $\mathbf{C}_2 \equiv \{e^{it} w\}$ two of such orbits, then $\mathbf{d}(e^{it} z, \mathbf{C}_2)$ does not depend on $t \in \mathbb{R}$. Indeed, since φ_t are isometries of $\mathbb{S}^3$ and $\varphi_t^{-1}(\mathbf{C}_2) = \mathbf{C}_2$, we have

$$\mathbf{d}(e^{it} z, \mathbf{C}_2) = \mathbf{d}(z, \varphi_t^{-1}(\mathbf{C}_2)) = \mathbf{d}(z, \mathbf{C}_2).$$

That property allows us to define the Clifford parallelism for great circles of $\mathbb{S}^3$.

Definition 2.1.1 Two great circles, $\mathbf{C}_1$ and $\mathbf{C}_2$, of $\mathbb{S}^3$ will be called Clifford parallel provided they are equidistant, that is, $\mathbf{d}(z, \mathbf{C}_2)$ does not depend on the point $z \in \mathbf{C}_1$. When that happens, we will write $\mathbf{C}_1 \| \mathbf{C}_2$.

Let us briefly study some properties related to the Clifford parallelism in $\mathbb{S}^3$. The main reference, as well as further details, will be the book by M. Berger [21].

Take a great circle $\mathbf{C}$ of $\mathbb{S}^3$ and a real number $\theta \in [0, \pi/2]$. Consider the set

$$\mathbf{T}_\theta = \{w \in \mathbb{S}^3 : \mathbf{d}(w, \mathbf{C}) = \theta\}.$$

For a better understanding of this set, we choose a coordinate system in $\mathbb{C}^2 \equiv \mathbb{R}^4$ as follows: let $\mathbf{P}$ be the plane containing $\mathbf{C}$ and set $\mathbf{P}^\perp$ the plane in $\mathbb{R}^4$ passing through the origin and orthogonal to $\mathbf{P}$. Then $\mathbf{C}^\perp$ is the great circle obtained when cutting $\mathbb{S}^3$ by the plane $\mathbf{P}^\perp$. Take now a coordinate system, $(z_1, z_2) \equiv (x_1, y_1, x_2, y_2)$, in $\mathbb{C}^2 \equiv \mathbb{R}^4$ adapted to the splitting $\mathbb{C}^2 \equiv \mathbb{R}^4 = \mathbf{P} \oplus \mathbf{P}^\perp$. Evidently $\mathbf{T}_0 = \mathbf{C}$ and $\mathbf{T}_{\pi/2} = \mathbf{C}^\perp$. For the remaining values of θ, the set $\mathbf{T}_\theta$ is nothing but a torus.

Proposition 2.1.2 *When $\theta \in (0, \pi/2)$, in the above coordinate system, $\mathbf{T}_\theta$ is the following rectangular torus*

$$\mathbf{T}_\theta = \{w = (w_1, w_2) \in \mathbb{S}^3 \subset \mathbb{C}^2 \,:\, |w_1| = \cos\theta,\ |w_2| = \sin\theta\}.$$

Proof In the chosen coordinate system set $\mathbf{C} = \{(z_1, 0) = (e^{it}, 0) \,:\, t \in \mathbb{R}\}$. Then $w = (w_1, w_2) \in \mathbf{T}_\theta$ means that $\mathbf{d}(w, \mathbf{C}) = \theta$, that is,

$$\mathbf{d}(w, \mathbf{C}) = \theta = \min_{z_1 \in \mathbf{C}} \{\arccos\left(\langle (w_1, w_2), (z_1, 0)\rangle\right)\} = \min_{z_1 \in \mathbf{C}} \{\arccos\left(\langle w_1, z_1\rangle\right)\}.$$

The arc reaches its minimum when the scalar product reaches its maximum. Indeed, varying z_1 in $\mathbf{C}$, its length does not change, its value is always one; w_1 is fixed, so that the variation goes with the cosine of the angle between z_1 and w_1. As a consequence, we have

$$\cos\theta = \max_{z_1 \in \mathbf{C}} \{\langle w_1, z_1\rangle\}.$$

That happens when z_1 and w_1 are collinear, which implies that

$$\cos\theta = |w_1|, \qquad |w_2|^2 = 1 - |w_1|^2 = \sin^2\theta.$$

□

Next we consider, in $\mathbb{C}^2 \equiv \mathbb{R}^4$, the following cone

$$\mathbf{Q}_\theta = \{w = (w_1, w_2) \in \mathbb{C}^2 \,:\, |w_1|^2 \sin^2\theta - |w_2|^2 \cos^2\theta = 0\}.$$

Note that the above torus is obtained when cutting $\mathbb{S}^3$ with that cone, i.e., $\mathbb{S}^3 \cap \mathbf{Q}_\theta = \mathbf{T}_\theta$. This fact is used (see [21] for a detailed proof) to show the following result.

Proposition 2.1.3 *Two different great circles which are Clifford parallel to* $\mathbf{C}$ *pass through each point* $w \in \mathbf{T}_\theta$.

Two important consequences should be pointed out:

(i) The Clifford parallelism of great circles of $\mathbb{S}^3$ is not an equivalence relation.
(ii) The corresponding geometric model associated to the Clifford parallelism does not satisfy the Euclidean fifth postulate.

Each great circle, $\mathbf{C}$, of $\mathbb{S}^3$ defines two subgroups, $\mathbf{G}^+$ and $\mathbf{G}^-$, of the group of isometries of $\mathbb{S}^3$ in such a way that the corresponding orbits when acting on $\mathbb{S}^3$ provide us the Clifford parallel great circles to the starting circle. They are usually called *first and second kind Clifford parallel*, respectively (see [21]). In the following section we will realize that by using the Hopf map and its conjugate.

2.2 The Hopf Map and Its Conjugate

We wish to identify the orbit space of the action (2.1). To do that let us consider the map

$$\Pi : \mathbb{S}^3 \to \mathbb{S}^2(1/2) \subset \mathbb{C}\times\mathbb{R}, \qquad \Pi(z_1, z_2) = \left(z_1\bar{z}_2, \frac{1}{2}(|z_1|^2 - |z_2|^2)\right), \tag{2.2}$$

where the image space is the 2-sphere $\mathbb{S}^2(1/2)$ of radius $1/2$ en $\mathbb{R}^3 \equiv \mathbb{C} \times \mathbb{R}$. This map can be viewed as a Riemannian submersion (see [22] and [82]). We are going to see that Π is just the quotient map associated to the action (2.1). In other words, *two points of* $\mathbb{S}^3$ *are in the same orbit of* (2.1) *if and only if they have the same image by the Hopf map* (2.2). That allows us to identify the orbit space of (2.1) with $\mathbb{S}^2(1/2)$. Indeed, if $z, w \in \mathbb{S}^3$ are in the same orbit, then $w = e^{it}z$, for a certain $t \in \mathbb{R}$, and so we get

$$w_1\bar{w}_2 = e^{it}z_1\, e^{-it}\bar{z}_2 = z_1\bar{z}_2.$$

Furthermore, $|z_1| = |w_1|$ and $|z_2| = |w_2|$, proving that $\Pi(z) = \Pi(w)$. Conversely, if $\Pi(z) = \Pi(w)$, then $|z_1|^2 - |z_2|^2 = |w_1|^2 - |w_2|^2$, and since $|z_1|^2 + |z_2|^2 = |w_1|^2 + |w_2|^2 = 1$, we obtain that $|z_1| = |w_1|$ and $|z_2| = |w_2|$. Write now $w_1 = e^{it_1}z_1$ and $w_2 = e^{it_2}z_2$, for $t_1, t_2 \in \mathbb{R}$. Finally, from $w_1\bar{w}_2 = z_1\bar{z}_2$ we find that

$$e^{it_1}z_1e^{-it_2}\bar{z}_2 = z_1\bar{z}_2 \quad\Longrightarrow\quad t_1 - t_2 = 0.$$

We will use the Hopf map to give a simple procedure to describe the Clifford parallel great circles of the first kind to a given great circle. The procedure is illustrated with the following algorithm:

1. Given a great circle $\mathbf{C}$ in $\mathbb{C}^2 \equiv \mathbb{R}^4$, take a coordinate system $z = (z_1, z_2)$ such that $z_2 = 0$ is the plane where $\mathbf{C}$ is lying. With that choice we can see $\mathbf{C}$ as the orbit of the point $\zeta = (1, 0) \in \mathbb{S}^3 \subset \mathbb{C}^2$ under the action (2.1), that is, $\mathbf{C} = \{(e^{it}, 0) : t \in \mathbb{R}\}$.
2. Then the great circles of the first kind Clifford parallel to $\mathbf{C}$ are obtained as the orbits of the action (2.1). More precisely, given a point $w = (w_1, w_2) \in \mathbb{S}^3 \subset \mathbb{C}^2 \equiv \mathbb{R}^4$, the great circle first kind Clifford parallel to $\mathbf{C}$ passing through w can be written as

$$\mathbf{C}_w = \{e^{it}w = (e^{it}w_1, e^{it}w_2 : t \in \mathbb{R}\}.$$

To describe the Clifford parallel great circles of the second kind we will use another action of $\mathbb{S}^1$ on $\mathbb{S}^3 \subset \mathbb{C}^2$ which will provide a new Riemannian submersion that will be called *conjugate Hopf map*. To do that let us define

$$\mathbb{S}^1 \times \mathbb{S}^3 \to \mathbb{S}^3, \qquad \varphi(e^{it}, z) = \varphi_t(z) = (e^{it} z_1, e^{-it} z_2). \tag{2.3}$$

To identify the orbit space we consider the following Riemannian submersion

$$\bar{\Pi} : \mathbb{S}^3 \to \mathbb{S}^2(1/2) \subset \mathbb{C}\times\mathbb{R}, \qquad \bar{\Pi}(z_1, z_2) = \left(z_1 z_2, \frac{1}{2}(|z_1|^2 - |z_2|^2) \right). \tag{2.4}$$

We see, as above, that two points are in the same orbit if and only if they have the same image by $\bar{\Pi}$. Furthermore, one can see that the orbits are Clifford parallel. Then we have a procedure to obtain Clifford parallel great circles passing through a given point:

1. Given a great circle $\mathbf{C}$ in $\mathbb{C}^2 \equiv \mathbb{R}^4$, take a coordinate system $z = (z_1, z_2)$ such that $z_2 = 0$ is the plane where $\mathbf{C}$ is lying. With that choice we can see $\mathbf{C}$ as the orbit of the point $\zeta = (1, 0) \in \mathbb{S}^3 \subset \mathbb{C}^2$ under the action (2.1), that is, $\mathbf{C} = \{(e^{it}, 0) : t \in \mathbb{R}\}$.
2. Then the great circles Clifford parallel to $\mathbf{C}$ appear as orbits of the actions (2.1) and (2.3). More precisely, given a point $w = (w_1, w_2) \in \mathbb{S}^3 \subset \mathbb{C}^2 \equiv \mathbb{R}^4$, the first and second kind great circles which are Clifford parallel to $\mathbf{C}$ passing through w can be written as

$$\mathbf{C}_w = \{(e^{it} w_1, e^{it} w_2) : t \in \mathbb{R}\}, \qquad \bar{\mathbf{C}}_w = \{(e^{it} w_1, e^{-it} w_2) : t \in \mathbb{R}\}.$$

It is worth noting that we can go from the first kind Clifford parallelism to the second one by means of the following isometry of $\mathbb{S}^3$

$$J : \mathbb{S}^3 \to \mathbb{S}^3, \qquad J(z_1, z_2) = (z_1, \bar{z}_2),$$

which makes commutative the following diagram

$$\begin{array}{ccc} \mathbb{S}^3 & \xrightarrow{J} & \mathbb{S}^3 \\ \Pi \downarrow & & \downarrow \bar{\Pi} \\ \mathbb{S}^2 & \xrightarrow{Id} & \mathbb{S}^2. \end{array}$$

Therefore, the Clifford parallelism splits in two equivalence relations, i.e., first kind and second kind Clifford parallelism. The isometry J is the mirror reflecting each other, so that it is enough to study only one of them.

2.3 Lancret Helices in Riemannian Spaces

We will approximate to the concept of Lancret helix in a 3-dimensional Riemannian space $(M, g = \langle , \rangle)$. Let $\gamma(t)$ be a curve in con M, with $t \in \mathbf{I} \subset \mathbb{R}$, and let $V(t)$ be a vector field along γ. Take a variation of γ in the direction of $V(t)$, that is, a map

$$\Gamma : \mathbf{I} \times (-\varepsilon, \varepsilon) \to M, \qquad (t, v) \mapsto \Gamma(t, v),$$

satisfying that $\Gamma(t, 0) = \gamma(t)$ and $\left(\frac{\partial \Gamma}{\partial v}\right)_{v=0} = V(t)$. Let us consider the geometric invariants of the longitudinal curves (z =constant) in the variation:

- Speed, $x(t, v) = |X(t, v)| = \left|\frac{\partial \Gamma}{\partial t}\right|$,
- Curvature, $\kappa(t, v)$,
- Torsion, $\tau(t, v)$,

and repeat the procedure we introduced in first chapter for curves in $\mathbb{R}^3$ to find the following variation formulas

$$V(x) = \langle \nabla_T V, T \rangle x, \tag{2.5}$$
$$V(\kappa^2) = 2\kappa \langle \nabla_T^2 V, N \rangle - 4\kappa^2 \langle \nabla_T V, T \rangle + \langle R(V, T)T, N \rangle, \tag{2.6}$$
$$\begin{aligned} V(\tau^2) = &-\frac{2\tau}{\kappa} \langle \nabla_T^3 V + R(\nabla_T V, T)T, B \rangle + \frac{2\tau\kappa'}{\kappa^2} \langle \nabla_T^2 V \\ &+ R(V, T)T, B \rangle - \langle \nabla_T V, \kappa B + \tau T \rangle. \end{aligned} \tag{2.7}$$

Assume now that $V(t)$ is obtained when a Killing field in $(M, g = \langle , \rangle)$, also called V, is restricted to the points of γ. It is well known that V generates a one-parametric group of isometries of $(M, g = \langle , \rangle)$, which will be denoted by $\{\mu_v : v \in \mathbb{R}\}$. Then we can define the following variation of $\gamma(t)$ in the direction of $V(t)$

$$\Gamma(t, v) = \mu_v(\gamma(t)).$$

As the speed, curvature and torsion are invariant by isometries, we obtain that

$$V(x) = V(\kappa^2) = V(\tau^2) = 0. \tag{2.8}$$

This fact yields the following concept, first introduced by J. Langer and D. Singer in [70]. *A vector field $V(t)$ along a curve $\gamma(t)$ will be called Killing along γ if it is a solution of* (2.8). This geometrically means that varying the curve $\gamma(t)$ in the direction of a Killing along the curve, then the shape of γ does not change, only its position. The following facts should be pointed out:

- If you have a curve with zero torsion ($\tau = 0$), then its unit binormal field, B, defines a Killing field along the curve.

- It is clear that the Eq. (2.8) form a system of linear equations whose solution space is 6-dimensional. That means that the space of Killing fields along any curve, in a 3-dimensional Riemannian space, is a 6-dimensional vector space.
- As we have just seen, any Killing field restricted to a curve provides a Killing field along the curve. Then we obtain a subspace, of the above vector space, whose dimension is that of the isometry group of $(M, g = \langle, \rangle)$.
- It is known that the dimension of the isometry group of a 3-dimensional Riemannian space is less than o equal to six, and equality holds for constant curvature spaces. As a consequence, in the latter spaces any Killing vector field along any curve is obtained when a Killing in the ambient space is restricted to the curve.

Remember that Lancret helices in $\mathbb{R}^3$ are curves making a constant angle with the flow generated by an infinitesimal translation, i.e., with a constant length Killing field in $\mathbb{R}^3$. This definition obviously holds good in any 3-dimensional Riemannian space. Nevertheless, it is not necessary to require that the length of the Killing field is constant everywhere, so that we propose a more general definition. *A curve $\gamma(t)$ in $(M, g = \langle, \rangle)$ will be called a Lancret helix if there exists a constant length Killing field, V, along the curve making a constant angle with the curve.* Then V will be called the axis of the Lancret helix.

2.4 Lancret Helices in Constant Curvature Spaces

In this section, whose main reference is [6], $(M, g = \langle, \rangle)$ will be a 3-dimensional Riemannian space of constant curvature c, where we aim to classify the Lancret helices.

As a first remark, we try to see if a zero torsion curve is automatically a Lancret helix. Indeed, $\tau = 0$ yields $\nabla_T B = 0$, so that $B(s)$ is a Killing field along $\gamma(s)$. Since we are working in a constant curvature space, any Killing field along $\gamma(s)$ comes from a Killing field in M. In this case, the constant angle, or the slope, is $\pi/2$. Furthermore, as it is well known, the curves with $\tau = 0$ are living in totally geodesic surfaces (and therefore they are of constant Gaussian curvature c) in M. These Lancret helices will be trivial and we are only interested in non trivial ones.

Let us start by analyzing the implications (necessary conditions) of an arc length parameterized curve $\gamma(s)$ to be a Lancret helix. We know that its axis is a constant length Killing field V, so that without loss of generality we can take of unit length in $(M, g = \langle, \rangle)$. Then we have

$$\langle T(s), V(s) \rangle = \langle \gamma'(s), V(s) \rangle = \cos\theta, \qquad \text{constant.} \tag{2.9}$$

Moreover, as $V(s)$ is Killing along $\gamma(s)$, the Eq. (2.8) hold, so that in (2.6) and (2.7) the curvature operator is $R(Z, T)T = cZ$, because $(M, g = \langle, \rangle)$ is of constant curvature c. Now $V(x) = 0$ yields $\langle \nabla_T V, T \rangle = 0$, which we combine with (2.9) to find that $\kappa\langle N, V\rangle = 0$. We can assume that $\kappa > 0$, otherwise we are working with a geodesic, which can be viewed as a trivial Lancret helix. Therefore we get $\langle N, V\rangle = 0$, which allows us to write

$$V(s) = \cos\theta\, T(s) + \sin\theta\, B(s).$$

As a consequence,

$$\nabla_T V(s) = (\kappa\cos\theta - \tau\sin\theta)N, \tag{2.10}$$

and

$$\begin{aligned}\nabla_T^2 V(s) &= -\kappa(\kappa\cos\theta - \tau\sin\theta)T + (\kappa'\cos\theta - \tau'\sin\theta)N\\ &\quad + \tau(\kappa\cos\theta - \tau\sin\theta)B.\end{aligned} \tag{2.11}$$

Now we combine (2.10) and (2.11) with (2.6) to obtain

$$\kappa(\kappa'\cos\theta - \tau'\sin\theta) = 0,$$

which implies that $\kappa\cos\theta - \tau\sin\theta$ should be a constant and so we write

$$\tau(s) = b\kappa(s) + a, \qquad b = \cot\theta. \tag{2.12}$$

To determine the constant a we use that $\nabla_T^2 V$ has no component in the direction of N, so that we have

$$\nabla_T^3 V = a\kappa'\sin\theta\, T - a\tau'\sin\theta\, B + a(\kappa^2 + \tau^2)\sin\theta\, N.$$

Bringing this information to (2.7) we obtain

$$\tau\sin\theta(a(\kappa\tau' - \tau\kappa') + c\kappa') = 0.$$

Since $\tau = 0$ yields trivial Lancret helices, we go to (2.12) to find

$$\kappa'(a^2 - c) = 0. \tag{2.13}$$

We conclude the following:

1. **Euclidean space**. If $c = 0$, we obtain Lancret's result, which has been widely discussed in the above chapter.
2. **Hyperbolic space**. If $c < 0$, then besides the trivial Lancret helices ($\tau = 0$), the only Lancret ones are those having both curvature and torsion constant. It is not difficult to see that in any constant curvature space the curves with constant curvature and torsion are automatically Lancret helices. Indeed, suffice it to take

the vector field $V(s) = \cos\theta\, T(s) + \sin\theta\, B(s)$, where $\cot\theta = \frac{\tau^2 - c}{\tau\kappa}$, and check that $V(s)$ is Killing along the curve. To show that it is enough to see that it satisfies the Eqs. (2.5), (2.6), and (2.7) in a space of constant curvature c. Then we know that there exists a Killing field in M that works as axis of the curve so that it becomes a Lancret helix. The following result has been shown:

Lancret theorem in the hyperbolic space: *A curve $\gamma(s)$ in $\mathbb{H}^3(-1)$ is a Lancret helix if and only if one of the following statements hols*

(i) $\tau = 0$ *and $\gamma(s)$ is a curve in some hyperbolic plane $\mathbb{H}^2(-1)$;*
(ii) *$\gamma(s)$ is a helix, i.e., a curve with curvature and torsion both constant.*

3. **Sphere**. If $c > 0$, we can assume that $c = 1$, that is, up to topology, we are in the unit 3-sphere $\mathbb{S}^3(1) \equiv \mathbb{S}^3$. In this case, the curvature and torsion of any non trivial Lancret helix are bound together by

$$\tau(s) = b\kappa(s) \pm 1, \qquad b = \cot\theta.$$

However, at first glance, it is not known whether this relation characterizes non trivial Lancret helices.

Surprisingly, that characterization holds. Indeed, let $\gamma(s)$ be a curve in $\mathbb{S}^3$ having curvature and torsion both constant and satisfying the above relation. Take θ with $\cot\theta = b$ to define the following vector field along $\gamma(s)$

$$V(s) = \cos\theta\, T(s) + \sin\theta\, B(s).$$

It is clear that it is of constant length and makes a constant angle, θ, with the curve. A straightforward computation allows us to find $\nabla_T V$, $\nabla_T^2 V$ and $\nabla_T^3 V$, so that we can check the Eqs. (2.5), (2.6), and (2.7). Therefore, we obtain a Killing field along the curve coming from a Killing field in $\mathbb{S}^3$. Summing up, we have shown the following result:

Lancret theorem in the sphere: *A curve $\gamma(s)$ in $\mathbb{S}^3$ is a Lancret helix if and only if one of the following statements holds*

(i) $\tau = 0$ and *$\gamma(s)$ is a curve in some 2-sphere $\mathbb{S}^2(1)$;*
(ii) *There exists a constant, the slope, b such that $\tau(s) = b\kappa(s) \pm 1$.*

Summarizing, it is worth noting the interest of the study of Lancret helices in the constant curvature spaces, mainly in Euclidean and spherical cases. Since we already made a detailed survey of such a curves in $\mathbb{R}^3$, it seems convenient to go further in considering Lancret helices in the 3-sphere.

2.5 The Inverse Problem and Natural Equations

In an n-dimensional Riemannian manifold it is known that its isometry group is a Lie group whose dimension is less than o equal to $\frac{1}{2}n(n+1)$. Furthermore, equality holds if and only if the space has constant sectional curvature. In this case we say that the space has the highest possible rigidity. Having the largest possible number of isometries can solve important problems in the context of what is known as *inverse problem*. In general, many structures (algebraic, geometric, analytic, etc.) arise when considering a mathematical model so that we have to define an equality relation (isomorphism, diffeomorphism, isometry, etc.) among them in order to study the properties of the chosen structure which remain invariant under the suitable relation. We are talking about the direct problem, that is, that concerning to find the so called *invariants* of the structure we are dealing with. Then the inverse problem consists of reproducing the structure from its invariants.

The best way to illustrate that situation is exhibiting an example which can be easily solved. Perhaps the simplest one is that concerning Linear Geometry, where two vector spaces which are isomorphic have the same dimension. The direct problem yields the dimension of a finite vector space is an invariant. The inverse problem consists of deciding if same dimension forces to be isomorphic. It is well known that we find an affirmative answer, so that the congruence classes of finitely generated vector spaces are in one-to-one correspondence with natural numbers.

As for the content of the present text, a problem of different nature should be stated. When studying regular curves in a 3-dimensional Riemannian space, we have a natural equality relation in the sense that two curves are congruent when they can be superposed, that is, if there is a rigid motion (isometry) that takes a curve on the other. There arise three geometric invariants concerning that congruence: length, curvature and torsion. We have to say they are of different nature. Indeed, whereas the length is intrinsic, the other two are extrinsic. This is an important fact, because the curves can be arc length parameterized, so that curvature and torsion remain as the very essential invariants. The inverse problem states as follows: *can we find an arc length parameterized curve from prescribed curvature and torsion functions*?

In the Euclidean space $\mathbb{R}^3$ you know an affirmative answer which is classically gathered in so called "fundamental theorem of space curves": *Let $\kappa(s) > 0$ and $\tau(s)$ be two functions defined on an interval, $I \subset \mathbb{R}$. Then there exists an arc length parameterized curve $\gamma(s)$, $s \in I$, in $\mathbb{R}^3$ with curvature function $\kappa(s)$ and torsion function $\tau(s)$. Furthermore, that curve is unique up to rigid motions in $\mathbb{R}^3$, that is, if we have two arc length parameterized curves with the same curvature and same torsion, there is a rigid motion that takes a curve on the other.*

- To show the first part, i.e., the existence of the curve, it is used that the natural equations of a curve constitute a system of linear differential equations which is governed by the curvature and torsion functions. The same idea can be considered in any 3-dimensional Riemannian space where there exist curves with prescribed curvature and torsion.

– When showing the uniqueness, up to motions, it is essential that the dimension of the isometry group of $\mathbb{R}^3$ is six, the largest possible dimension. The same proof can be done for 3-dimensional Riemannian spaces with 6-dimensional isometry group. They are the constant curvature spaces, such as $\mathbb{S}^3$ and $\mathbb{H}^3$. Spaces less rigid, those of non constant curvature, do not possess many isometries, so we can find non congruent curves having same curvature and torsion.

Consequently, the inverse problem for curves can be solved in any 3-dimensional space. But knowing the existence of a curve with prescribed curvature and torsion is only a theoretical fact. We have to go further, that is, to find an explicit equation of the curve, a problem classically known *solving natural equations*.

A few examples will clarify the situation. As for plane curves we have that *any plane curve can be recovered from its curvature function by quadratures*. Then, given a function $\kappa(s)$, need not be positive, we wish to determine the curve that, up to plane motions, has that function as its curvature. To do that let us consider the function $\theta(s)$ which, at least locally, determine the angle of the curve with the x-axis. Then we know that

$$\theta(s) = \int_0^s \kappa(v)\, dv,$$

so that the indicatrix tangent of the curve is

$$T(s) = \left(\cos \int_0^s \kappa(v)\, dv, \sin \int_0^s \kappa(v)\, dv \right).$$

A new integration allows us to obtain the curve $\gamma(s)$ arc length parameterized, where $\gamma'(s) = T(s)$.

It can be also observed that, when studying Lancret helices in $\mathbb{R}^3$, we met the *solving natural equations* problem for such a curves. Indeed, we had success in getting a geometrical integration of the natural equations of Lancret helices based on the property that they are geodesics of right cylinders over plane curves, and then reducing the problem to plane curves. That can be summarized in the following algorithm:

1. We wish to find, up to motions in $\mathbb{R}^3$, the Lancret helix with curvature $\kappa(s)$ and torsion $\tau(s)$, satisfying that $\dfrac{\tau(s)}{\kappa(s)} = \cot\theta$.
2. Let us start by considering, for instance, in the plane $z = 0$, the arc length parameterized curve $\alpha(s)$, with curvature function $\kappa_\alpha(s) = \dfrac{\kappa(s)}{\sin^2\theta}$.
3. Now let $\mathbf{C}_\alpha = \{\alpha(s) + z\partial_z\, s \in \mathbf{I},\ z \in \mathbb{R}\}$ be the right cylinder over α, which is naturally a flat surface.
4. On that cylinder, take the geodesic $\beta(u) = \alpha(\sin\theta\, u) + \cos\theta\, u\, \partial_z$, with slope θ, which is a Lancret helix with curvature $\kappa(s)$ and torsion $\tau(s)$.

2.6 The Hopf Map as a Riemannian Submersion

Now we are going to study the *solving natural equations* for Lancret helices in $\mathbb{S}^3$. The key tool will be the Hopf map (2.2). As we said before, this map is a Riemannian submersion, so that we can define the vertical distribution $\mathcal{V}_z = \mathrm{Ker}(d\Pi_z)$. It is a 1-dimensional integrable distribution, whose leaves are precisely the orbits of action (2.1) and therefore it is globally generated by the infinitesimal translation $V(z) = iz$. The orthogonal distribution is the horizontal one $\mathcal{H}_z = (\mathcal{V}_z)^\perp$, so we have that

$$T_z\mathbb{S}^3 = \mathcal{H}_z \oplus \mathcal{V}_z, \quad \text{orthogonal splitting.}$$

Furthermore, $d\Pi_z$ provides a linear isometry between $\mathcal{H}_z$ and $T_{\Pi(z)}\mathbb{S}^2$.

It should be observed that the orthogonal complement of $\mathcal{H}_z$ in $\mathbb{C}_z \equiv \mathbb{R}^4_z$ is the plane generated by $\{z, iz\}$, so that when $x \in \mathcal{H}_z$ we have $ix \in \mathcal{H}_z$.

A very useful tool to study Riemannian submersions is the technique of horizontal liftings, which adapted to our case works as follows. Given a curve $\alpha : [0, a) \to \mathbb{S}^2$ with $\alpha(0) = p \in \mathbb{S}^2$, then for each $z \in \Pi^{-1}(p)$ there exists a unique curve $\bar{\alpha} : [0, a) \to \mathbb{S}^3$ satisfying the following conditions

(i) $\bar{\alpha}(0) = z$;
(ii) $\Pi(\bar{\alpha}(t)) = \alpha(t)$, for any $t \in [0, a)$, which means that $\bar{\alpha}$ is a lifting of α;
(iii) The curve $\bar{\alpha}$ is horizontal, that is, $\bar{\alpha}'(t) \in \mathcal{H}_{\bar{\alpha}(t)}$.

Henceforth, by *horizontal lifting* we will mean curves obtained as above. It is clear that there exists a horizontal lifting for any initial condition. Therefore, if $\bar{\alpha}(s)$ is a horizontal lifting of $\alpha(s)$, then all of curves $\beta_t(s) = e^{it}\,\bar{\alpha}(s)$, $t \in \mathbb{R}$ are also horizontal liftings of α. Moreover, all horizontal liftings are obtained in this way. Note that when $\alpha(s)$ is arc length parameterized, then all of its horizontal liftings also are, because $d\Pi$ is a linear isometry on the horizontal planes.

The horizontal liftings of curves can be extended to to vector fields. Indeed, given X a vector field on an open set U of $\mathbb{S}^2$, it can be defined a vector field $\bar{X}$ on the open $\Pi^{-1}(U)$ of $\mathbb{S}^3$ which is horizontal and obviously Π-related to X, which means that

(i) $d\Pi(\bar{X}) = X$ along $\Pi^{-1}(U)$,
(ii) $\bar{X}_z \in \mathcal{H}_z$, for any $z \in \Pi^{-1}(U)$.

As a consequence, we can talk about horizontal liftings of vector fields, which usually will be written with overbars, as well as horizontal liftings of any geometrical object in the base space $\mathbb{S}^2$.

In the study of isometric immersions (submanifolds) will appear the so called fundamental equations, the first ones known as Gauss and Weingarten, the former regarding the variation of tangent vectors fields to the sub manifold, the latter for normal vector fields. As for Riemannian submersions the corresponding fundamental equations are known as O'Neill equations ([81] and [82]). Formally, they deal with the horizontal and vertical parts in the four possible combinations when differentiate horizontal and vertical vector fields. In general, the equations are governed by two invariants, O'Neill's invariants. Geometrically, the first one measures what the leaves of the submersion are far from being totally geodesics. The other measures what the horizontal distribution is far from being integrable. In our case the computations are quite simple so that it will not be necessary to resort to those invariants.

Let $\overline{\nabla}$ be the Riemannian connection of $\mathbb{S}^3$ and let ∇ be the connection of the base space $\mathbb{S}^2$. Observe first that in any Riemannian submersion any vector field of the form $[\bar{X}, V]$ is vertical. This is a consequence of the Π-relation, that is,

$$d\Pi([\bar{X}, V]) = [d\Pi(\bar{X}), d\Pi(V)] = [X, 0] = 0.$$

In this context, the O'Neill equations for the Hopf map are

$$\overline{\nabla}_{\bar{X}} \bar{Y} = \overline{\nabla_X Y} - \langle i\bar{X}, \bar{Y}\rangle V, \tag{2.14}$$

$$\overline{\nabla}_{\bar{X}} V = \overline{\nabla}_V \bar{X} = i\bar{X}, \tag{2.15}$$

$$\overline{\nabla}_V V = 0. \tag{2.16}$$

Before to prove the equations, it should be noted that the third one (2.16) means that the orbits are geodesics. This property should be already known, because the orbit passing trough $z \in \mathbb{S}^3$ is the curve

$$\delta(t) = e^{it} z = \cos t\, z + \sin t\, iz,$$

which is the great circle obtained when cutting $\mathbb{S}^3$ by the plane generated by $\{z, iz\}$, and so it is a geodesic.

To show those equations will be essential that V is a Killing field in $\mathbb{S}^3$, which is equivalent (see any basic Riemannian Geometry textbook) to say that $\overline{\nabla} V$ is antisymmetric, that is,

$$\langle \overline{\nabla}_A V, B\rangle = -\langle \overline{\nabla}_B V, A\rangle.$$

Then an alternative proof of (2.16) is the following. It is clear that $\overline{\nabla}_V V$ has no vertical component, since V is an infinitesimal translation (it has constant length). As for the horizontal one we have that

$$\langle \overline{\nabla}_V V, \bar{X}\rangle = -\langle \overline{\nabla}_{\bar{X}} V, V\rangle = -\frac{1}{2}\bar{X}(|V|^2) = 0.$$

To show (2.15) we will see that $[\bar{X}, V] = 0$. Indeed, as it is vertical we have that

$$\langle [\bar{X}, V], V\rangle = \langle \overline{\nabla}_{\bar{X}} V, V\rangle - \langle \overline{\nabla}_V \bar{X}, V\rangle = \frac{1}{2}\bar{X}(|V|^2) + \langle \overline{\nabla}_V V, \bar{X}\rangle = 0.$$

Then we obtain the first part of (2.15). Furthermore, it is clear that

$$\overline{\nabla}_{\bar{X}} V = \overline{\nabla}_{\bar{X}} iz = i \overline{\nabla}_{\bar{X}} z = i\bar{X},$$

which completes the proof of (2.15). Finally, to prove (2.14) we bear in mind that, in any Riemannian submersion the horizontal component of $\overline{\nabla}_{\bar{X}} \bar{Y}$ is precisely $\overline{\nabla_X Y}$, so that it is enough to compute the vertical part, which is given by

$$\langle \overline{\nabla}_{\bar{X}} \bar{Y}, V \rangle = -\langle \overline{\nabla}_{\bar{X}} V, \bar{Y} \rangle = -\langle i\bar{X}, \bar{Y} \rangle.$$

2.7 Hopf Tubes

Given $\beta(s)$ a curve in $\mathbb{S}^2$, which can be taken arc length parameterized and defined on the whole $\mathbb{R}$, we consider its preimage, or total lifting by the Hopf map. Then we obtain all of horizontal liftings of the curve

$$\Pi^{-1}(\beta) = \{e^{it}\,\overline{\beta}(s) \;:\; (s,t) \in \mathbb{R}^2\},$$

where $\bar{\beta}$ is a horizontal lifting of β. We have found a surface in $\mathbb{S}^3$ which can be parameterized by means of the following covering map

$$X : \mathbb{R}^2 \to \Pi^{-1}(\beta) \subset \mathbb{S}^3, \qquad X(s,t) = e^{it}\,\bar{\beta}(s).$$

There we see the two families of coordinate curves

- The curves t =constant are horizontal liftings of $\beta(s)$;
- The curves s =constant are the orbits of the action (2.1), in other words, the integral curves of V.

Compute the induced metric, in this parameterization, to obtain

$$X_s = \frac{\partial X}{\partial s} = e^{it}\,\bar{\beta}'(s), \qquad X_t = \frac{\partial X}{\partial t} = ie^{it}\,\bar{\beta}(s) = V(X(s,t)),$$

$$g_{11} = \langle X_s, X_s \rangle = 1, \quad g_{12} = \langle X_s, X_t \rangle = 0, \quad g_{22} = \langle X_t, X_t \rangle = 1.$$

That implies that the Gaussian curvature vanishes, that is, $\Pi^{-1}(\beta)$ is a flat surface, which will be called the *Hopf tube*, and denoted by $\mathbf{T}_\beta$, built over $\beta(s)$, which works as the transversal section of that tube. The following properties are easy to check:

1. The Hopf tubes, in general, are immersed in $\mathbb{S}^3$ and they are embedded if and only if the curve $\beta(s)$ is a simple (without selfintersections) curve in $\mathbb{S}^2$.
2. If $\beta(s)$ is closed, then $\mathbf{T}_\beta$ is a flat torus (*Hopf torus*) whose isometry type will be computed later.

3. It is also clear that Hopf tubes are surfaces which are invariantes under the action (2.1), so that they can be seen generated by the orbits of that action directed by a horizontal lifting of the transversal section. In other words, its generatrices are Clifford parallel great circles of the first kind (second kind when working with (2.4)). In this sense, they are the counterpart of the right cylinders in the Euclidean geometry.
4. The converse also holds. Any surface in $\mathbb{S}^3$ invariant under the action (2.1) is a Hopf tube over a certain curve in $\mathbb{S}^2$. Indeed, if S is a surface invariant under (2.1), then the orbit passing through any point of S is contained in the surface in such a way that S is foliated by orbits. Then we can consider a curve in S being orthogonal at a point to the orbit passing through it. In other words, we take a leaf in S of the foliation orthogonal to the vertical one. Such a curve is obviously horizontal and so a horizontal lifting of its projection by (2.2). It is clear that S agrees with the Hopf tube built over the projected curve.

2.8 Lancret Helices and Hopf Cylinders in $\mathbb{S}^3$

We have seen that Lancret helices in $\mathbb{R}^3$ can be viewed as geodesic of circular cylinders, and conversely. In what follows we will see an analogous result for Lancret helices in $\mathbb{S}^3$.

To do that, we will begin by recalling the Pinkall [86] description of the Hopf map $h : \mathbb{S}^3(1) \to \mathbb{S}^2(1/2)$. Here $\mathbb{S}^n(r)$ will represent the n-dimensional sphere of radius $r > 0$ and $\mathbb{S}^2(1/2)$ will be identified with the 1-dimensional complex projective space $\mathbb{CP}^1$ with constant holomorphic sectional curvature 4. Similarly, $\mathbb{S}^3(1)$ will be identified with the unit quaternions $\{q \in \mathbf{H} : q\bar{q} = 1\}$ and then $\mathbb{S}^2(1/2)$ will be the 2-sphere of radius 1/2 in the subspace of $\mathbf{H}$ generated by $\{1, j, k\}$. Let $q \mapsto \bar{q}$ be the antiautomorphism of $\mathbf{H}$ such that $i \mapsto -i$ and fixes $\{1, j, k\}$. Then $h(q) = \frac{1}{2}\bar{q}\,q$ (see [86] for the properties of h). In particular, h is a Riemannian submersion, so we can use O'Neill's notation and terminology [82]. Therefore, given $p \in \mathbb{S}^3(1)$, the tangent space $T_p\mathbb{S}^3(1)$ decomposes in the horizontal plane (which is isometric, via dh, to $T_{h(p)}\mathbb{S}^2(1/2)$) and a vertical line that is the tangent space to fiber over p.

Let $\overline{\nabla}$ and ∇ be the Levi-Civita connections of $\mathbb{S}^3(1)$ and $\mathbb{S}^2(1/2)$, respectively, and write the horizontal liftings from the base space by overbars objects. We then have the following formulas

$$\begin{aligned} \overline{\nabla}_{\overline{X}}\overline{Y} &= \overline{\nabla_X Y} - ((\langle JX, Y\rangle \circ h)V, \\ \overline{\nabla}_{\overline{X}}V &= \overline{\nabla}_V\overline{X} = i\overline{X}, \\ \overline{\nabla}_V V &= 0, \end{aligned} \tag{2.17}$$

where V is a unit vertical vector field and J represents the standard complex structure of $\mathbb{S}^2(1/2)$. Note that the last equation expresses the geodesic nature of the fibers.

Let $\beta : I \subset \mathbb{R} \to \mathbb{S}^2(1/2)$ be a curve of unit speed with Frenet frame $\{\beta', \xi\}$ and curvature function λ. Let us take a horizontal lifting $\bar{\beta}$ of β with Frenet frame $\{\bar{\beta}', \xi^*, \eta^*\}$ and curvature functions λ^* and μ^*. We combine all three previous equations with those of Frenet of β and β^* to deduce that $\lambda^* = \lambda \circ h = \bar{\lambda}$ and $\xi^* = \bar{\xi}$. In particular, ξ^* belongs to the horizontal distribution along β. It is not difficult to see that $\mu^* = \pm 1$ and that $\eta^* = \pm V$, that is, the binormal η^* of $\bar{\beta}$ and the unit tangent to the fibers at each point of $\bar{\beta}$ coincide.

Through $h^{-1}(\beta)$ we obtain the total lifting of β, $M_\beta \subset \mathbb{S}^3(1)$, which we know to be a flat surface of $\mathbb{S}^3(1)$ (see [86]). We will call M_β *the Hopf cylinder over* β, which admits a parameterization $X : I \times \mathbb{R} \to \mathbb{S}^3(1)$ of the form

$$X(t, z) = e^{iz}\bar{\beta}(t) = \cos z\, \bar{\beta}(t) + \sin z\, V(t).$$

Note that in this parameterization the t-curves are the horizontal liftings of β, while the z-curves correspond to the fibers. Both families of curves are parameterized by the arc length and are mutually orthogonal. Furthermore, they are geodesics of M_β. Pinkall also showed that the Hopf cylinder M_α associated to a closed curve α of length L, enclosing an oriented area A in $\mathbb{S}^2(1/2)$, is isometric to the flat torus $\mathbb{R}^2/\Gamma$, where Γ is the lattice generated by $(2A, L)$ and $(2\pi, 0)$. Let us also highlight that holonomy prevents that horizontal liftings of β are closed.

The Weingarten map $\mathcal{A}$ of M_β into $\mathbb{S}^3(1)$ is given by

$$\begin{aligned} \mathcal{A}(X_t) &= \bar{\lambda} X_t + X_z, \\ \mathcal{A}(X_z) &= X_t \end{aligned} \tag{2.18}$$

so we present the following result:

Theorem 2.8.1 *Let β be a curve embedded in $\mathbb{S}^2(1/2)$ and M_β its Hopf cylinder in $\mathbb{S}^3(1)$. Then each geodesic of M_β is a Lancret helix that lives entirely in $\mathbb{S}^3(1)$.*

Proof Let $\alpha(s)$ be a geodesic of M_β. Then exists a constant φ (the slope) such that

$$T(s) = \alpha'(s) = \cos\varphi\, X_z + \sin\varphi\, X_t. \tag{2.19}$$

This equation proves that T makes a constant angle φ with X_z, which is a constant length Killing vector field along α, which completes the proof. □

The converse result is stated as follows:

Theorem 2.8.2 *Every Lancret helix that lives entirely in $\mathbb{S}^3(1)$ can be seen as a geodesic of a certain Hopf cylinder in $\mathbb{S}^3(1)$.*

Proof A Lancret helix γ, with curvature κ and torsion τ, satisfies that $\tau = b\kappa \pm 1$, where $b = \cot\theta$, and θ is the angle between the axis of the helix and γ'. We define

β, up to motions in $\mathbb{S}^2(1/2)$, as the curve whose curvature function is $\lambda = (1 + b^2)\kappa \pm 2b$. Then in M_β choose the geodesic α associated to the slope φ defined as $\cot\varphi = \mp b$. It is easy to see that γ and α have the same curvatures, and therefore they are congruent in $\mathbb{S}^3(1)$. □

Finally, for closed curves and their corresponding Hopf tori we have

Corollary 2.8.3 *Let β be a curve embedded in $\mathbb{S}^2(1/2)$. There is a rational one-parameter family of closed Lancret helices in $\mathbb{S}^3(1)$ living in the Hopf torus M_β.*

Proof If L denotes the length of β and A the area that it encloses in $\mathbb{S}^2(1/2)$, then M_β is isometric to the flat torus obtained from the lattice

$$\Lambda = \{(2A, L), (2\pi, 0)\}.$$

As we have seen, the Lancret helices living in M_β correspond to the geodesics of M_β. Let γ be a geodesic of M_β, which as before we will denote by φ its slope. Then γ is closed if and only if there exists an $s_0 > 0$ such that $\gamma(s_0) = \gamma(0)$. The latter occurs if and only if $(s_0 \cos\varphi, s_0 \sin\varphi) \in \Lambda$. Hence, there exists a rational number q such that

$$\cot\varphi = \frac{2A}{L} + q\frac{2\pi}{L}. \tag{2.20}$$

□

Chapter 3
The Conformal Extension and Protein Chains

3.1 The Essential Conformal Extension

The model of Lancret helix in $\mathbb{S}^3$ that has been just described holds for many helicoidal configurations, but it does not cover all of them. The following considerations will help to clarify the situation (see [7]).

(i) The first detail concerns the axis of the helix, that is, to the discussion that we did about the axis of the helix being a Killing vector field, either translational or rotational. In this sense, we know that circular helices (Lancret helices with curvature and torsion both constant) admit as a second axis a rotational Killing vector field perpendicular to the axis of the cylinder on which is coiled and tangent to the cross section. The role played for both axes is completely different. Indeed, it is true that Lancret helices are loxodromes on cylinders over plane curves, a property that arises from the existence of the first axis. The existence of a rotational Killing field as a second axis does not imply that the helix wraps around a cylinder on a plane curve.

(ii) We have already seen that Lancret helices in $\mathbb{R}^3$ do not close, since its tangent field always has a positive component, just the one in the direction of the axis. Now, it have known for a long time cyclic peptides, as the gramycin S antibiotic and the immunosuppressive agent cyclospirin, on the contrary to circular proteins that have only been known for a decade. In recent years, a large number of circular proteins in bacteria, animals and plants have been discovered. In that class the so-called round or cyclotid proteins are included (see for example [25]). At a first glance, it seems natural to think that circular protein chains can be modeled by means of some type of helix with a circular axis.

To describe both circular proteins and other more complicated helical configurations of chains of proteins, we propose a conformal variational model, which we will develop based on the following ideas:

A. Ferrandez Izquierdo, *Helices*, RSME Springer Series 16,
https://doi.org/10.1007/978-3-032-09663-0_3

1. The fact that there are no closed Lancret helices in the model described for $\mathbb{R}^3$ is due, in part, to the fact that the straight lines integrals of the field that acts as axis are not closed. So that such a flow of parallel lines to be closed, we need to change the ambient space topology. In this sense, it seems natural to choose the one-point compactification, that is, go to $\mathbb{S}^3 = \mathbb{R}^3 \cup \{\zeta_o\}$. Let us say we have added the point at infinity.
2. Next, $\mathbb{S}^3$ will be endowed with a metric as similar as possible to the Euclidean one of $\mathbb{R}^3$. That is not difficult for it suffices to choose one in the conformal class in order to preserve the angles. Furthermore, we have metrics with constant curvature which allow us to see $\mathbb{S}^3$ as a round sphere.
3. In this context, "straight lines" are geodesics, that is, great circles. Moreover, we have a classic structure that allows us to talk about parallel lines: the so-called Clifford parallelism. The flows of Clifford parallel great circles are very well described by Hopf vector fields, namely, infinitesimal translations that can be viewed as vertical flows of Hopf maps. Thus, we have a kind of cylinders or tubes in $\mathbb{S}^3$, playing the role of circular cylinders, which are obtained by lifting curves, by means of the Hopf map, in $\mathbb{S}^2$ that come to be cross sections of tubes. Then the Lancret helices in $\mathbb{S}^3$ can be characterized of Lancret, that is to say, they are the curves that admit as axis an infinitesimal translation, or a Hopf vector field, as the geodesics of a Hopf tube.
4. But there is more, because we found the Lancret helices of $\mathbb{S}^3$ as critical points of an energy action that depends on the corresponding geometric invariants. We also get the full family of closed Lancret configurations. In other words, the so-called *problem of the closed curve* for Lancret helices in $\mathbb{S}^3$.
5. Once that problem is solved, it is conformally projected on $\mathbb{R}^3$. To do so, we will use the stereographic projection from the added point ζ_o. The Hopf flow is transformed into a Villarceau flow, so that conformal helices form a constant angle with this flow, the axis being a conformal Killing vector field. They can be seen as loxodromes in conformal Hopf tubes, including the Clifford torus. Now we can also project stereographically from any point, with the corresponding deformation of the surfaces and helices, which now appear as rhumb lines in Dupin cyclids.
6. Let us say that the underlying idea in the new geometric framework that we propose is to see the flow of Villarceau circles, which is the conformal image of the Clifford parallelism in $\mathbb{S}^3$, as substitute of the flow of parallel straight lines in the Euclidean sense. So, the model provides solutions that are helicoidal structures whose axis is already a flow of parallel lines in the Euclidean sense, and a flow of parallel Villarceau circles in the Clifford sense.

Therefore, we choose a 3-sphere with a constant curvature metric, i.e. a round sphere, which can be assumed, without loss of generality, of radius one, which we will denoted simply by $\mathbb{S}^3$. Next, we will detail the program that we have just presented.

3.2 Lancret Helices in $\mathbb{S}^3$ Revisited

Once the general ideas have been established to define the model that we pursue and the tool of Clifford parallelism, we propose the following program:

- Obtain geometrically the Lancret helices of $\mathbb{S}^3$, that is, use geometry to solve the natural equations of the Lancret helices in $\mathbb{S}^3$.
- Characterize the Lancret helices in $\mathbb{S}^3$ as solutions of a variational problem associated to an action that only depends on the geometric invariants of the trajectories.
- Find an explicit algorithm to characterize the closed Lancret helices in $\mathbb{S}^3$, that is, to solve the closed curve problem relative to Lancret helices in $\mathbb{S}^3$.

Analytic Approach Lancret helices in $\mathbb{S}^3$, as in $\mathbb{R}^3$, admits a nice characterization in terms of their geometric invariants, curvature and torsion (see [6]). We have just seen that a curve γ in $\mathbb{S}^3$ is a Lancret helix if, and only if, it satisfies one of the following two conditions:

1. $\tau = 0$ and so γ is in a totally geodesic 2-round sphere $\mathbb{S}^2$;
2. The curvature and torsion of γ are subject to the condition

$$\tau = b\,\kappa \pm 1, \qquad \text{for some constant } b.$$

In general, if you work on a round sphere of radius $R > 0$, the above equation writes down as $\tau = b\,\kappa \pm 1/R$.

Geometric Approach We know that Lancret helices in $\mathbb{S}^3$ can also be characterized as geodesics of certain flat tubes. On the other hand, the one-parameter group $\mathbf{G}_C^+ = \{\varphi_t \,:\, t \in \mathbb{R}\}$ generates the following field in $\mathbb{S}^3$

$$V(\zeta) = \left\{ \frac{d}{dt}\left(\varphi_t(\zeta)\right) \right\}_{t=0} = i\,\zeta, \quad \forall \zeta \in \mathbb{S}^3,$$

which is a Killing vector field of constant length that defines the vertical flow of the Riemannian immersion. These types of vector fields on $\mathbb{S}^3$ are usually called *Hopf vector fields*.

Given a curve $\delta : I \subset \mathbb{R} \to \mathbb{S}^2(1/2)$ in a round sphere of radius $1/2$, its full lifting $\mathbf{T}_\delta = h^{-1}(\delta)$ is a flat surface in $\mathbb{S}^3$, the Hopf tube over δ or Hopf tube with cross section δ. This surface is parameterized as

$$X : I \times \mathbb{R} \to \mathbf{T}_\delta \subset \mathbb{S}^3, \quad X(s,t) = e^{it}\,\bar{\delta}(s),$$

where $\bar{\delta}$ is a horizontal lifting of δ. So the coordinates curves are, respectively, δ-horizontal liftings (t = constant) and orbits or fibers (s = constant). The following result (see [6]) proves, in particular, that the natural equations for Lancret helices in $\mathbb{S}^3$ can be integrate by quadratures (see [7]).

Theorem 3.2.1 *A curve γ in $\mathbb{S}^3$ is a Lancret helix if and only if, up to motions of $\mathbb{S}^3$, it is a geodesic of a Hopf tube.*

Proof Note, first, that the curvature and torsion of a geodesic γ of a Hopf tube $\mathbf{T}_\delta$ are given by

$$\kappa = \frac{\kappa_\delta + 2b}{1+b^2}, \qquad \tau = \frac{-1 + b\kappa_\delta + b^2}{1+b^2},$$

where κ_δ is the curvature function of δ in $\mathbb{S}^2(1/2)$ and b denotes the slope, measured with respect to the fibers, of γ as a geodesic of the flat surface $\mathbf{T}_\delta$. This implies that γ is a Lancret helix in $\mathbb{S}^3$.

Conversely, if γ is a Lancret helix in $\mathbb{S}^3$, we have that $\tau = b\kappa \pm 1$ for some constant b. Let us now consider a curve δ in $\mathbb{S}^2(1/2)$, with curvature $\kappa_\delta = (1 + b^2)\kappa \pm 2b$ (this curve is unique, up to movements in the 2-sphere). Now, in the Hopf tube $\mathbf{T}_\delta$, we choose the geodesic γ_b, making an angle φ with the fibers, where $\cot\varphi = \mp b$ (the geodesic with slope b). It is not difficult to see that γ and γ_b have the same curvature and torsion functions and thus, up to parameterization, they are congruent in $\mathbb{S}^3$. □

Corollary 3.2.2 *Up to motions in $\mathbb{S}^3$, the Lancret helices are described by one of two equivalent moduli:*

1. *(κ, τ): The curvature and torsion functions, which must satisfy a well known constraint.*
2. *(κ_δ, b): The curvature function in $\mathbb{S}^2(1/2)$ of the cross section δ and the slope as a geodesic of $\mathbf{T}_\delta$.*

In any case, the moduli space of Lancret helices can be identify with the space $C^\infty(\mathbb{R}) \times \mathbb{R}$.

Variational Approach To complete the cycle, the Lancret helices in $\mathbb{S}^3$ can be characterized as solutions of a variational principle. Indeed, take $q_1, q_2 \in \mathbb{S}^3(R)$ (the 3-sphere of radius R) and let $\{\vec{x}_1, \vec{y}_1\}$, $\{\vec{x}_2, \vec{y}_2\}$ be two orthonormal vectors in $T_{q_1}\mathbb{S}^3(R)$ and $T_{q_2}\mathbb{S}^3(R)$, respectively. Consider the space of clamped curves

$$\Lambda = \{\gamma : [t_1, t_2] \to \mathbb{S}^3(R) \,:\, \gamma(t_i) = q_i,\ T(t_i) = \vec{x}_i,\ N(t_i) = \vec{y}_i,\ 1 \le i \le 2\}.$$

In this space we have the following triparametric family of functionals $\{\mathcal{F}_{mnp} : \Lambda \to \mathbb{R} \,:\, m, n, p \in \mathbb{R}\}$, defined by

$$\mathcal{F}_{mnp}(\alpha) = \int_\alpha (m + n\kappa + p\tau)\, ds. \tag{3.1}$$

The field equations associated to these Lagrangians, for curves in $\mathbb{S}^3$ and using a standard method (see [4]), are

$$-m\kappa + p\kappa\tau + n\left(\frac{1-R^2\tau^2}{R^2}\right) = 0,$$

$$(p\kappa - n\tau)' = 0,$$

The moduli spaces of solutions are represented in the following table, where, for simplicity, different cases have been distinguished according to the values of the parameters.

m	n	p	Moduli space of configurations in $\mathbb{S}^3(R)$
$\neq 0$	$= 0$	$= 0$	Geodesics $\kappa = 0$
$= 0$	$= 0$	$\neq 0$	Circles (κ constant and $\tau = 0$)
$= 0$	$\neq 0$	$= 0$	Horizontal liftings, via Hopf map, of curves in $\mathbb{S}^2$
$\neq 0$	$\neq 0$	$= 0$	Helices with $\kappa = \frac{n(1-R^2\tau^2)}{mR^2}$
$\neq 0$	$= 0$	$\neq 0$	Helices with any κ and $\tau = \frac{m}{p}$
$= 0$	$\neq 0$	$\neq 0$	Helices with $\kappa = \frac{n(1-a^2R^2)}{apR^2}$, $\tau = \frac{1}{aR^2}$ and $a \in \mathbb{R} - \{0\}$
$\neq 0$	$\neq 0$	$\neq 0$	Helices with $\kappa = \frac{n(1-a^2R^2}{(m+ap)R^2}$, $\tau = \frac{p+maR^2}{(m+ap)R^2}$ and $a \in \mathbb{R} - \{-\frac{m}{p}\}$
$\neq 0$	$\neq 0$	$\neq 0$	Lancret helices de Lancret with $\tau = \frac{p}{n}\kappa - \frac{m}{p}$ and $R = \pm\frac{p}{m}$

As a consequence, we have the following variational approach for Lancret helices in $\mathbb{S}^3$ (see [7]).

Theorem 3.2.3 *A curve in $\mathbb{S}^3$ is a Lancret helix if, and only if it is a critical point of some action $\mathcal{F}_{mnp}$ acting on a suitable space of curves.*

Quantization Principle for Closed Lancret Helices To finish the Lancret program in $\mathbb{S}^3$, we need a description of closed Lancret helices. This is achieved with the following argument, which culminates in a principle of quantization.

- As Lancret helices are geodesics of Hopf tubes, which are flat surfaces, we need to start from a closed curve δ in $\mathbb{S}^2(1/2)$. If the cross section is closed, then it is not difficult to see that $\mathbf{T}_\delta$ is a torus. To determine the isometry group of this flat torus, we consider the covering map

$$X : \mathbb{R}^2 \to \mathbf{T}_\delta, \quad X(s,t) = e^{it}\,\bar{\delta}(s),$$

and known machinery is used (see [50] and [86]) to find that $\mathbf{T}_\delta$ is isometric to $\mathbb{R}^2/\Gamma$, where Γ is the lattice in the Euclidean plane generated by $(L, 2A)$ and (0.2π). Here $L > 0$ denotes the length of δ and $A \in (-\pi, \pi)$ is the oriented area bounded by δ in the 2-sphere.

- As a consequence, a Lancret helix in $\mathbb{S}^3$ is closed if and only if its slope h satisfies the following quantization condition

$$h = \frac{1}{L}(2A + q\pi), \quad q \in \mathbb{Q} \text{ rational.}$$

- The existence of closed Lancret helices in any Hopf torus is guaranteed by the isoperimetric inequality in $\mathbb{S}^2(1/2)$. Indeed, the length and the area enclosed by embedded closed curve are related by

$$L^2 + 4A^2 - 4\pi A \geq 0,$$

which can be written as

$$L^2 + (2A - \pi)^2 \geq \pi^2.$$

Therefore, in the plane $(L, 2A)$, we define the region

$$\Delta = \{(L, 2A) \, : \, L^2 + (2A - \pi)^2 \geq \pi^2 \, \text{and} \, 0 \leq A \leq \pi\}.$$

Thus, for any point $a = (L, 2A) \in \Delta$ there exists a closed embedded curve δ^a in $\mathbb{S}^2(1/2)$ of length L and enclosed area A. Now we compare with the previous quantization principle of the slope to see that the geodesic of slope h in the torus of $\mathbf{T}_\delta$ closes if and only if the straight line in the $(L, 2A)$-plane with slope h intersects the axis $2A$ at a height that is a rational multiple of π. Therefore, the moduli space of closed Lancret helices in $\mathbb{S}^3$ is identified with the following region of the plane (see Fig. 3.1)

$$\Delta \cap \left(\bigcup_{q \in \mathbb{Q}} \left(\frac{p}{n} L - 2A = q\pi \right) \right).$$

3.3 Examples

We will now give a general procedure for constructing Hopf tubes in $\mathbb{S}^3 \subset \mathbb{C}^2$. First of all we observe that the Hopf map can be written in the form

$$h : \mathbb{S}^3 \subset \mathbb{C}^2 \to \mathbb{S}^2(1/2) \subset \mathbb{C} \times \mathbb{R}, \quad h(z_1, z_2) = \left(z_1 \bar{z}_2 \, , \, \frac{1}{2}(|z_1|^2 - |z_2|^2) \right).$$

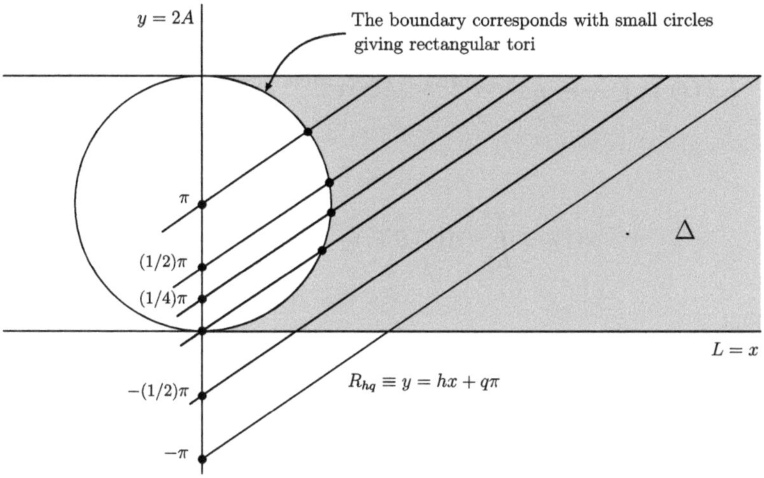

Fig. 3.1 Closed critical curves for a given slope. The intersection points with y-axis are rational multiples of π

Given a point $p = (a + ib, c) \in \mathbb{S}^2(1/2)$, its fiber in $\mathbb{S}^3$ is given by

$$h^{-1}(p) = \left\{ \left(\frac{\sqrt{1+2c}}{\sqrt{2}} e^{it}, \frac{\sqrt{2}}{\sqrt{1+2c}} (ae^{it} + be^{i(t-\pi/2)}) \right) : t \in \mathbb{R} \right\}.$$

Let $\delta : I \subset \mathbb{R} \to \mathbb{S}^2(1/2)$ be a curve of the form $\delta(s) = (a(s), b(s), c(s))$. Then its Hopf tube is the set

$$\begin{aligned}\mathbf{T}_\delta &= h^{-1}(\delta)\\ &= \left\{ \left(\frac{\sqrt{1+2c(s)}}{\sqrt{2}} e^{it}, \frac{\sqrt{2}}{\sqrt{1+2c(s)}} (a(s)e^{it} + b(s)e^{i(t-\pi/2)}) \right) : s \in I,\ t \in \mathbb{R} \right\}.\end{aligned}$$

In order to parameterize $\mathbf{T}_\delta = h^{-1}(\delta)$ using fibers and horizontal liftings of the transversal section δ, we have to determine a horizontal lifting $\bar{\delta}$ of δ. To do that, let us put $t = \alpha(s)$ in the above formula and determine $\alpha(s)$ using the horizontality condition $\langle i\bar{\delta}(s), \bar{\delta}'(s)\rangle = 0$. We get

$$\alpha'(s) = B'(s)A(s) - A'(s)B(s),$$

where

$$A(s) = \frac{\sqrt{2}\,a(s)}{\sqrt{1+2c(s)}}, \quad B(s) = \frac{\sqrt{2}\,b(s)}{\sqrt{1+2c(s)}}.$$

Then

$$\bar{\delta}(s) = \left(\frac{\sqrt{1+2c(s)}}{\sqrt{2}} e^{i\alpha(s)}\,,\; A(s)\, e^{i\alpha(s)} + B(s) e^{i(\alpha(s)-\pi/2)} \right),$$

where

$$\alpha(s) = \int_{s_o}^{s} (A(s)B'(s) - A'(s)B(s))\, ds.$$

So that

$$\mathbf{T}_\delta = h^{-1}(\delta) \equiv X(s,t) = e^{it} \left(\frac{\sqrt{1+2c(s)}}{\sqrt{2}} e^{i\alpha(s)}\,,\; A(s)\, e^{i\alpha(s)} + B(s) e^{i(\alpha(s)-\pi/2)} \right).$$

Let us look at some explicit examples of Hopf tubes.

Clifford Torus Seen as a Hopf Tube Maybe the most popular compact surface, at least of genus one, of $\mathbb{S}^3$ is the so-called Clifford torus. It appears in many classification problems of surfaces as well as in some interesting and old conjecture. The simplest way to define the Clifford torus is given by the map

$$Y : \mathbb{R}^2 \to \mathbb{C}^2, \quad Y(u,v) = \frac{\sqrt{2}}{2}(e^{i\sqrt{2}u}, e^{i\sqrt{2}v}). \tag{3.2}$$

It is clear that $Y(\mathbb{R}^2) \subset \mathbb{S}^3$. Moreover, this map is biperiodic with period $\sqrt{2}\,\pi$ and, therefore, it defines an embedding, which we will also denote by Y, of the squared torus or Riemannian product of two circles of radii $\sqrt{2}/2$ in $\mathbb{S}^3$, that is,

$$Y : \mathbb{S}^1(\sqrt{2}/2) \times \mathbb{S}^1(\sqrt{2}/2) \to \mathbb{S}^3, \quad Y(u,v) = \frac{\sqrt{2}}{2}(e^{i\sqrt{2}u}, e^{i\sqrt{2}v}),$$

is the well known Clifford torus.

But now we want to see the Clifford torus as a Hopf tube whose cross section is a geodesic of $\mathbb{S}^2(1/2)$. Therefore, we choose

$$\delta : \mathbb{R} \to \mathbb{S}^2(1/2), \quad \delta(s) = \left(\frac{1}{2} \cos 2s, \frac{1}{2} \sin 2s, 0 \right).$$

We will make use of the above standard argument to see that $\alpha(s) = s$ and so

$$\bar{\delta} : \mathbb{R} \to \mathbb{S}^3, \quad \bar{\delta}(s) = \frac{\sqrt{2}}{2} \left(e^{is}, e^{-is} \right),$$

it is a horizontal lifting. Thus, the Clifford torus can be parameterize using horizontal liftings (t = constant) and fibers (s = constant) as follows:

$$X : \mathbb{R}^2/\Gamma \to \mathbb{S}^3 \subset \mathbb{C}^2, \quad X(s,t) = \frac{\sqrt{2}}{2}\left(e^{i(s+t)}, e^{i(-s+t)}\right),$$

where Γ is the lattice, in $\mathbb{R}^2$, generated by (π, π) and $(0, 2\pi)$.

It is evident that $X(s,t) = Y\left(\frac{1}{\sqrt{2}}(s+t), \frac{1}{\sqrt{2}}(-s+t)\right)$ and so it we get

$$X_s = \frac{1}{\sqrt{2}}Y_u - \frac{1}{\sqrt{2}}Y_v, \quad X_t = \frac{1}{\sqrt{2}}Y_u + \frac{1}{\sqrt{2}}Y_v.$$

Since the curves u = constant and v = constant are geodesics of the Clifford torus, they are Lancret helices in $\mathbb{S}^3$ with slope $h = 1$ and $h = -1$, respectively (in this case, they are circular helices, since δ has constant curvature).

Rectangular Tori Such as Hopf Tubes Whose Sections Transversals Are Geodesic Circles The above example can be extend to rectangular tori in $\mathbb{S}^3$, which can be seen as Hopf tubes over $\mathbb{S}^2(1/2)$ geodesic circles. Those tori can be defined as follows:

$$Y : \mathbb{S}^1(r_1) \times \mathbb{S}^1(r_2) \to \mathbb{S}^3, \quad Y(u,v) = (r_1\, e^{iu/r_1}, r_2\, e^{iv/r_2}),$$

with $r_1^2 + r_2^2 = 1$. As before, the coordinate curves u = constant and v = constant are geodesics of those tori $Y(\mathbb{S}^1(r_1) \times \mathbb{S}^1(r_2)) \subset \mathbb{S}^3$.

On the other hand, it is considered a geodesic circle of $\mathbb{S}^2(1/2)$, that is, a small circle, say

$$\delta : \mathbb{R} \to \mathbb{S}^2(1/2), \quad \delta(s) = \left(r \cos\frac{s}{r}, r \sin\frac{s}{r}, m\right), \quad r^2 + m^2 = \frac{1}{4}.$$

In this case, we see that $\alpha(s) = \lambda\, s$ with $\lambda = \sqrt{\frac{1-2m}{1+2m}}$, and so the curve

$$\bar{\delta}(s) = \left(\sqrt{\frac{1+2m}{2}}\, e^{i\lambda s}, \sqrt{\frac{1-2m}{2}}\, e^{i(\lambda - 1/r)s}\right)$$

is a horizontal lifting of δ.

Therefore, the rectangular torus parameterized as a tube of Hopf is given by

$$X : \mathbb{R}^2/\Gamma \to \mathbb{S}^3 \subset \mathbb{C}^2, \; X(s,t) = \left(\sqrt{\frac{1+2m}{2}}\, e^{i(\lambda s+t)}, \sqrt{\frac{1-2m}{2}}\, e^{i((\lambda-1/r)s+t)}\right),$$

where Γ is the lattice in $\mathbb{R}^2$ generated by

- $(\pi\sqrt{1-4m^2}\,, \pi(1-2m))$ and $(0, 2\pi)$, if $m > 0$, or
- $(\pi\sqrt{1-4m^2}\,, \pi(1+2m))$ and $(0, 2\pi)$, if $m < 0$.

It is clear that $r_1 = \sqrt{\frac{1+2m}{2}}$, $r_2 = \sqrt{\frac{1-2m}{2}}$, and

$$X(s,t) = Y\left(r_1(\lambda s + t), r_2((\lambda - 1/r)s + t)\right),$$

so that

$$X_s = \lambda r_1 Y_u + r_2(\lambda - 1/r)Y_v, \qquad X_t = r_1 Y_u + r_2 Y_v.$$

Since the curves u = constant and v = constant are geodesics of the rectangular torus, then they are Lancret helices in $\mathbb{S}^3$ (in this case they are circular helices, since δ has constant curvature).

The Hopf Tube Whose Cross Section is a Viviani Curve In 1692, Vincenzo Viviani (1622–1703), a disciple of Galileo, challenged European mathematicians to solve a problem of the quadrature of a part of the surface of a sphere. In *Aenigma geometricum de miro opificio testudinis quadrabilis hemisphaericae (The geometric problem of finding the area of a hemisphere intersected with a cylinder)* he asked how four equal windows could be cut from a hemisphere so that the remaining surface can be exactly squared. His solution involved using the intersection of four right cylinders, the bases of which are tangent to the base of the hemisphere.

The answer to such a question involves the Viviani curve. Such a curve, seen in $\mathbb{S}^2(1/2)$, is obtained by cutting that sphere with the right cylinder $(x-1/4)^2+y^2 = 1/16$. This is how you get the curve

$$\delta(s) = \left(\frac{1+\cos s}{4}, \frac{\sin s}{4}, \frac{\sin\frac{s}{2}}{2}\right),$$

which is closed with period 4π, for example in $-2\pi \le s \le 2\pi$. Note that this curve is not parameterized by the arc, however, in the previous argument that leads us to the Hopf tube over a cross section, the parameterization does not matter. As a consequence, we can follow, step by step, the above argument to get

$$\alpha'(s) = \frac{1}{4}\left(1 - \sin\frac{s}{2}\right), \quad \alpha(s) = \frac{1}{4}\left(s + 2\cos\frac{s}{2}\right).$$

A horizontal lifting of the Viviani curve turns out to be

$$\bar{\delta}(s) = \left(\frac{\sqrt{1+\sin\frac{s}{2}}}{\sqrt{2}}\, e^{i\alpha(s)}, \frac{\sqrt{2}}{4\sqrt{1+\sin\frac{s}{2}}}\left(e^{i\alpha(s)} + e^{i(\alpha(s)-s)}\right)\right).$$

This leads to a good parameterization of the Hopf torus, over the Viviani curve, of the form

$$X : [-2\pi, 2\pi] \times \mathbb{R} \rightarrow \mathbf{T}_\delta \subset \mathbb{S}^3 \subset \mathbb{C}^2,$$

given by

$$X(s,t) = \left(\frac{\sqrt{1+\sin\frac{s}{2}}}{\sqrt{2}} e^{i(\alpha(s)+t)}, \frac{\sqrt{2}}{4\sqrt{1+\sin\frac{s}{2}}} \left(e^{i(\alpha(s)+t)} + e^{i(\alpha(s)-s+t)} \right) \right).$$

3.4 Villarceau Flows: A Conformal Field Theory to Describe Protein Chains

If $\mathbf{T}$ is a torus of revolution in $\mathbb{R}^3$, it is known that $\mathbf{T}$ contains two families of circles, the parallel and the meridians. However, it is less well known that $\mathbf{T}$ also contains another type of circles, called *Villarceau circles*, as they were discovered by A. J. Yvon Villarceau (1813–1883) in 1848. The Villarceau circles in $\mathbf{T}$ appear as the intersection of $\mathbf{T}$ with a bitangent plane. Thus, two families are found, $\mathcal{F}_1 = \{\Upsilon(t)\}$ and $\mathcal{F}_2 = \{\Xi(t)\}$ of those exotic circles. Two circles of different families intersect at exactly two points, while two of the same family do not intersect, but are linked like two links in a chain.

There is a nice relationship between Clifford parallel great circles and Villarceau circles through an adequate stereographic projection. Let us take $\zeta_o \in \mathbb{S}^3 \subset \mathbb{R}^4$ and consider the stereographic projection $E_o : \mathbb{S}^3 \setminus \{\zeta_o\} \rightarrow \mathbb{R}^3$ which, as it is known, is the restriction of an inversion in $\mathbb{R}^4$ with pole ζ_o. Now let C be a great circle passing through ζ_o. Let us choose a coordinate system $\{x, y, z\}$ in $\mathbb{R}^3$ such that the z-axis is $E_o(C \setminus \{\zeta_o\})$ and then $E_o(C^\perp)$ will be the unit circle of the plane $\{x, y\}$.

In this framework, it is not difficult to see that $\mathbf{T}_\theta = E_o(C_\theta)$, $\theta \in (0, \frac{\pi}{2})$, is a torus of revolution around $E_o(C \setminus \{\zeta_o\})$ in $\mathbb{R}^3$. Furthermore, up to similarities, each revolution torus in $\mathbb{R}^3$ is of the form $E_o(C_\theta)$, for a suitable value $\theta \in (0, \frac{\pi}{2})$. The two families of Villarceau circles in $\mathbf{T}_\theta = E_o(C_\theta)$ are obtained as images by the stereographic projection of E_o of the two types of great circles in C_θ that are Clifford parallel to C. So, all Villarceau circles in $\mathbb{R}^3 \setminus$(axis Z) are described as follows:

$$\mathcal{F}_1 = \{\Upsilon(t)\} = \{E_o(C') : C' \text{ is of 1st kind Clifford parallel to } C\},$$

and

$$\mathcal{F}_2 = \{\Xi(t)\} = \{E_o(C'') : C'' \text{ is of 2nd kind Clifford parallel to } C\}.$$

In what follows, we will refer to these circles as *Villarceau circles of the first and second kind*, depending on whether they are in $\mathcal{F}_1$ or in $\mathcal{F}_2$, respectively. Observe that $\mathcal{F}_1$ (respectively, $\mathcal{F}_2$) defines a foliation of $\mathbb{R}^3 \backslash (z - \text{axis})$.

Note that in $\mathcal{F}_1$ and in $\mathcal{F}_2$ we have included a circle that is neither Villarceau circle first kind nor second kind in a torus of revolution around the z-axis. Such a circle is $E_o(C^{\perp})$, which, however, will also be considered as one of Villarceau.

The Clifford parallelism is projected onto $\mathbb{R}^3$ and thus can be described in terms of Villarceau circles. Namely, as C is the orbit through $\zeta_o \in C$ and we have chosen $E_o(C)$ as the z-axis in $\mathbb{R}^3$, we have a group of conformal applications that preserves the orientation of $\mathbb{R}^3 \setminus (z - \text{axis})$ associated to C, and defined by

$$\mathbf{H}_{\mathbf{C}}^{+} = E_o \circ \mathbf{G}_C^{+} \circ E_o^{-1} = \{\psi_t = E_o \circ \varphi_t \circ E_o^{-1} \ : t \in \mathbb{R}\}.$$

Thus, the orbits in $\mathbb{R}^3 \setminus (z - \text{axis})$ associated to $\mathbf{H}_C^{+}$ are exactly the Villarceau circles of the first kind on a family of torus of revolution around the z-axis.

Note that given a pair of Villarceau circles of the first kind, say γ_1 and γ_2, then $\gamma_1 = E_o(C_1)$ and $\gamma_2 = E_o(C_2)$ for certain great circles that satisfy $C \,||^{+}\, C_1$, $C \,||^{+}\, C_2$, so $C_1 \,||^{+}\, C_2$. In other words, those Villarceau circles are images, via a stereographic projection, of Hopf fibers. However, they are lying either on the same revolution torus, or on different revolution torus. The first occurs when $d(C, C_1) = d(C, C_2)$ and the second when $d(C, C_1) \neq d(C, C_2)$.

Similarly, one can consider

$$\mathbf{H}_C^{-} = E_o \circ \mathbf{G}_C^{-} \circ E_o^{-1} = \{E_o \circ \chi_t \circ E_o^{-1} \ : t \in \mathbb{R}\}$$

the group of conformal maps that leave invariant the Villarceau circles of second kind on a family of revolution torus around the z-axis.

From now on, we will restrict ourselves to Villarceau flows of the first kind, although anything would be worth if we worked with those of second kind. Therefore, once we have solved the variational problems whose solutions are Lancret helices in $\mathbb{S}^3$, we project, via a stereographic map, to get helices in $\mathbb{R}^3$ whose axis is a conformal field (which generates a one-parameter group of conformal transformations on $\mathbb{R}^3$). More explicitly, and without loss of generality, let us consider $\zeta_o = (0, i) \in \mathbb{S}^3 \subset \mathbb{C}^2$. So, in the described situation, the stereographic projection with pole ζ_o is given by

$$E_o : \mathbb{S}^3 \backslash \{\zeta_o\} \to \mathbb{R}^3 \equiv \mathbb{C} \times \mathbb{R}, \quad E_o(z_1, z_1) = \left(\frac{z_1}{1 - \mathrm{Im}(z_2)}, \frac{\mathrm{Re}(z_2)}{1 - \mathrm{Im}(z_2)} \right).$$

Now, let us choose, for example, a Clifford torus $\mathbf{T}_{\text{Clifford}} \equiv Y(u, v)$, parameterized as a Riemannian product of circles (3.2). It is not hard to see that its stereographic image $E_o(\mathbf{T}_{\text{Clifford}})$, is a revolution torus in $\mathbb{R}^3$ whose meridians are the curves $E_o(u = \text{constant})$ and the parallels are $E_o(v = \text{constant})$.

On the other hand, that torus, viewed as a Hopf tube $\mathbf{T}_{\text{Clifford}} \equiv X(s,t)$ is projected onto the conformally flat surface $E_o(\mathbf{T}_{\text{Clifford}})$, proving that it is foliated by Villarceau circles of the first kind, which are the images by E_o of the fibers s = constant. Now, the Lancret helices in $\mathbf{T}_{\text{Clifford}}$ form a constant angle with the fibers, and therefore with the curves v = constant. Since E_o is a conformal mapping, the stereographic projection of a Lancret helix makes a constant angle with the parallels of the torus $E_o(\mathbf{T}_{\text{Clifford}})$. Choosing, in particular, closed Lancret helices in the Clifford torus and projecting them by E_o, we obtain beautiful closed configurations that are good candidates for modeling circular protein chains.

But there is more, since what we have done with Clifford torus also holds for rectangular tori and, in general, for any Hopf tube (in particular, any Hopf torus). Indeed, the key point of this discussion is a conformal field in $\mathbb{S}^3$. It is the corresponding stereographic image of the Hopf vector field $V(\zeta) = i\,\zeta$ that governs the theory of Lancret helices in $\mathbb{S}^3$. Therefore, we define in $\mathbb{R}^3\backslash(z-\text{axis})$ the following vector field

$$W \in \chi(\mathbb{R}^3\backslash(z-\text{axis})), \qquad W = dE_o(V).$$

It produces the Villarceau flow $\mathcal{F}_1$. Alternatively, it is a conformal vector field that generates the one-parameter group of conformal transformations $\mathbf{H}_C^+$.

The Villarceau flows of the first kind, that is, the flows of W in $\mathbb{R}^3\backslash(z-\text{axis})$, can be explicitly obtained as follows. Note that any Villarceau circle cuts the half-plane $P = \{(x,0,z) : x > 0\}$ exactly at a point and remember that two Villarceau circles of the same type do not intersect. Then, for any $p = (x,0,z) \in P$, the Villarceau circle of the first kind $\gamma_p : [-\pi,\pi] \to \mathbb{R}^3$, which passes through p, is given by

$$\gamma_p(t) = E_o\left(e^{it}x_1, e^{it}((x_2 + iy_2)\right),$$

where $E_o^{-1}(x,0,z) = (x_1,0,x_2,y_2)$, and therefore

$$\gamma_p(t) = \frac{1}{1 - x_2\sin t - y_2\cos t}(x_1\cos t, x_1\sin t, x_2\cos t - y_2\sin t).$$

The length and radius of this circle are, respectively,

$$L = \frac{1 + \|p\|^2}{x}\pi, \qquad r = \frac{1 + \|p\|^2}{2x}.$$

Now, given a regular curve δ in $\mathbb{S}^2(1/2)$, consider its Hopf tube $\mathbf{T}_\delta$, which we know is a flat surface in $\mathbb{S}^3$. Then we project it, via the stereographic map, to obtain a surface $\mathbf{M}_\delta = E_o(\mathbf{T}_\delta) \subset \mathbb{R}^3$, which we call *conformal Hopf tube with conformal cross section* δ. These surfaces are tangent to $W = dE_o(V)$ and are thus foliated by

Villarceau circles, namely,

$$\mathbf{M}_\delta = E_o(\mathbf{T}_\delta) = \bigcup_{p \in \mathbf{M}_\delta} E_o\left(C_{E_o^{-1}(p)}\right),$$

where $C_{E_o^{-1}(p)}$ is the fiber in $\mathbb{S}^3$ over $E_o^{-1}(p)$.

A Lancret helix γ in $\mathbb{S}^3$ is a geodesic in some Hopf tube $\mathbf{T}_\delta$, and forms a constant angle with the Hopf vector field V, which is a Killing vector field on the 3-sphere. As a consequence, its stereographic image $\beta = E_o(\gamma)$ (we will call these curves *conformal Lancret helices*) is in the conformal Hopf tube $\mathbf{M}_\delta = E_o(\mathbf{T}_\delta)$ and makes a constant angle with the vector field $W = dE_o(V)$. In other words, the conformal Lancret helices lie in conformal Hopf tubes and make a constant angle with the Villarceau circles.

In the model that we propose to describe the helical configurations in Nature, the solutions are conformal Lancret helices. They are conformal images in $\mathbb{R}^3$ of curves in $\mathbb{S}^3$ that are critical points of an action whose Lagrangian density is linear in the curvature and torsion functions. This discussion culminates in the following result, which from a geometric point of view can be seen as a conformal Lancret integration theorem (see [7]).

Theorem 3.4.1 *A curve β in $\mathbb{R}^3$ is a conformal Lancret helix if and only if it makes a constant angle with the Hopf vector field $W = dE_o(V)$. Furthermore, conformal Lancret helices lie in conformal Hopf tubes and form a constant angle with the Villarceau circles.*

Remark 3.4.2

(i) It should be noted that when the Hopf vector field comes from the second Hopf map, h_-, the the helices associated to Villarceau flow are of the second kind.
(ii) It can also be projected from another point of $\mathbb{S}^3$, say ζ. Then the helical structures appear as loxodromes, with respect to a conformal vector field, lying in conformal Hopf tubes which are deformed as Dupin cyclides.

Chapter 4
Lancret Helices in Homogeneous 3-Spaces

According to a classical result of Whitney and Grauestein [109], the total curvature of plane curves is, under suitable boundary conditions, an invariant in the homotopy class of the curve. Even more, each homotopy class of curves, satisfying suitable boundary conditions, is completely characterized by the common total curvature. As an obvious consequence, if fluctuations do not change the topology of curves, then a variational approach to the total curvature action of curves in the Euclidean plane turns out to be trivial.

On the other hand, another classical and well known result, due to Fenchel [41], assures that the total curvature of any simple closed curve in the Euclidean 3-space satisfies

$$\int_{\gamma} \kappa(s)\, ds \geq 2\pi,$$

with equality if and only if γ is a convex plane curve. Therefore, the minimum of the total curvature action on simple closed curves in the Euclidean space is 2π and it is reached just on the convex plane curves. Hence, an obvious and naive question is to decide whether, besides those minima, there is any other extremal (critical) of the total curvature functional acting on closed curves in the Euclidean space.

More precisely, it seems natural to consider the total curvature functional acting on a suitable space of curves in a certain surface, or more generally in a Riemannian space, and then to study the associated variational problem. Although the study of the total curvature of curves in Riemannian spaces has been intensively considered along the literature (see [31] and references therein), the systematic study of the associated variational approach was initiated in [2] and [3]. However, we have to point out that M. S. Plyushchay, [87], proposed it as a model to study massless particles with rigidity.

Some progress has been made in this direction which, as far as we know, can be briefly summarized as follows. The variational approach for the total curvature in surfaces was first considered in [3], where it was shown that extremals of the total

A. Ferrandez Izquierdo, *Helices*, RSME Springer Series 16,
https://doi.org/10.1007/978-3-032-09663-0_4

curvature are reached by curves consisting of parabolic points. Stability of extremals was also holographically characterized there. Then, the problem in spaces with the highest rigidity (constant curvature) was considered in [2], where it was shown that the dynamics associated with the total curvature action is consistent only in round 3-spheres. More precisely, when studying the total curvature dynamics of curves in an n-dimensional Riemannian space, $M^n(k)$, with constant curvature k, we have:

- The dynamics is reduced to dimension $n \leq 3$.
- The curvature must be non-negative, $k \geq 0$, so it does not make sense in, for example, the hyperbolic space.
- For $k = 0$ and up to topology, the dynamics actually occurs over arbitrary plane curves in $\mathbb{R}^3$. This, in particular, gives the answer to the above stated naive question.
- For $k > 0$, up to topology, the extremals over a round 3-sphere are just the so called Legendrian curves, that is, curves which are horizontal liftings, via the Hopf map, of curves in the 2-sphere.

Other partial results, providing examples and families of extremals in Berger spheres and in the complex projective plane, can be found in [2] (see also [15], [26] and [29] for extremals in warped product spaces).

We will focus on this variational approach for curves in homogeneous 3-spaces whose isometry group has dimension four (sometimes called rotationally symmetric homogeneous 3-spaces [64]). In particular, we obtain the complete classification of extremals for the total curvature action in these backgrounds (see [16]). The main tool to determine the extremals is the existence of an infinitesimal translation, ξ, which allows us to see each homogeneous 3-space with rigidity of order four as a fibration over a surface with constant curvature, which works, in turn, as the space of ξ-orbits. As a consequence, a congruence class of these homogeneous 3-spaces is determined, up to topology, by a pair of constants: the curvature of the base, c, and the so called bundle curvature, r. Then, following [34], we will use the popular notation $\mathrm{E}(c, r)$, with $c \neq 4r^2$, to denote those homogeneous 3-spaces having four dimensional isometry group (if $c = 4r^2$ the space has constant curvature, in other words, highest rigidity, and six dimensional isometry group). Furthermore, we provide algorithms to explicitly construct the extremals, which can be regarded as the solution of the so called solving natural equations for extremals problem. Our results provide progress in the knowledge of the dynamics associated with the total curvature action of curves in 3-dimensional homogeneous geometries (including Thurston ones [99] and [100]), or, in other words using the Plyushchay terminology [87], of the dynamics of a massless boson in those backgrounds. The current status of this variational approach can be now described as follows:

1. In the hyperbolic space, $\mathbb{H}^3$, there is no dynamics of curves associated with the total curvature action, because it does not provide any extremal.

2. If $c = 4r^2$, then $\mathrm{E}(c, r)$ has constant curvature, r^2, and, up to topology, it should be either
   2.1 $\mathbb{R}^3$, when $r = 0$, and the dynamics works through curves with vanishing torsion, i.e., plane curves; or
   2.2 $\mathbb{S}^3(r^2)$, the round sphere with curvature r^2, when $r \neq 0$. Now, the dynamics works through curves with torsion $\pm r$. In other words, horizontal liftings via the Hopf map of curves in the corresponding round two-sphere.
3. If $c \neq 4r^2$ and $r = 0$, then the homogeneous 3-space is a Riemannian product and there are two possibilities:
   3.1 If $c > 0$, then $\mathrm{E}(c, 0) = \mathbb{S}^2(c) \times \mathbb{R}$ and there are two families of extremals for the total curvature action, namely,
      3.1.1 The class of curves with horizontal Frenet normal. It forms a real one-parameter class $\{\gamma_m, m \in \mathbb{R} - \{0\}\}$ of Lancret helices, that is, curves making a constant angle with ξ. More precisely, this family can be described as follows: for any $m \in \mathbb{R} - \{0\}$, choose in $\mathbb{S}^2(c)$ the circle β_m with geodesic curvature

$$\kappa_g = \frac{\sqrt{c(m^2+1)}}{m},$$

      and then take γ_m as the geodesic with slope m in the Hopf cylinder shaped on β_m.
      3.1.2 The class of curves with horizontal binormal, which automatically have torsion zero. Up to rigid motions, this family is formed by curves lying in Hopf cylinders built over great circles in $\mathbb{S}^2(c)$ (see Corollary 4.3.11).
   3.2 If $c < 0$, then $\mathrm{E}(c, 0) = \mathbb{H}^2(c) \times \mathbb{R}$. Now the space of extremals is made up of curves with horizontal binormal. Up to rigid motions, they are the curves lying in Hopf cylinders built over geodesics in the hyperbolic plane $\mathbb{H}^2(c)$ (see Corollary 4.3.11).
4. If $c \neq 4r^2$ and $r \neq 0$, the space of extremals consists of two families of curves:
   4.1 A real one-parameter class of Lancret helices (see Corollary 4.3.4).
   4.2 The class of curves with horizontal binormal, which automatically has constant torsion $\tau = \pm r$. This class can be parameterized by the space of differentiable real functions. They are summarized as well as explicitly and geometrically described after Proposition 4.3.2.

Besides the isotropic and the rotationally symmetric homogeneous spaces, whose isometry groups have, respectively, dimension six and four, a third class completes the list of 3-dimensional homogeneous spaces (see [92], [99], and [100]). Homogeneous spaces in this class have an isometry group of dimension three, the corresponding geometries are called solv-geometries or Bianchi geometries and they are associated with the group Sol_3. Therefore, a natural open problem, which will not be addressed here, is to study the dynamics associated with the total curvature of curves in Sol_3.

Our interest, then, is directed towards the closed curve problem. In other words, we wonder about the existence of closed extremals, that is, critical points of the total curvature energy acting on closed curves in $\mathrm{E}(c, r)$. We completely solve this question showing two big families of closed solutions:

(i) A first class, $\mathbf{F}_N$, found in any $\mathrm{E}(c, r)$ other than $\mathrm{H}^2(c) \times \mathbb{S}^1$, is formed by a rational one-parameter family of closed Lancret helices, equivalently, curves with horizontal normal.
(ii) A second class, $\mathbf{F}_B$, which can be found in any $\mathrm{E}(c, r)$, is formed by closed solutions with horizontal binormal. This family is identified with the set of convex closed curves in the Euclidean plane.

Moreover, we give a quantization principle, in the sense of Dirac, which assures that the critical levels of the total curvature energy acting on closed curves in $\mathrm{E}(c, r)$ are not arbitrary, but rational multiples of a certain fixed quantity of energy (see [16]).

4.1 The Total Curvature Action

Let us consider clamped curves, with a finite number of inflection points, in an n-dimensional Riemannian space $(M, \langle\, , \rangle)$. More precisely, for any pair of points, $p_1, p_2 \in M$ and tangent vectors $x_i \in T_{p_i}M$, we consider the space, Ω, of smooth immersed curves with a finite number of points where the curvature vanishes (inflection points) that are clamped to those points, that is,

$$\delta : [a_1, a_2] \to M \qquad \delta(a_i) = p_i, \;\; \delta'(a_i) = x_i, \;\; 1 \leq i \leq 2.$$

Note that this space includes, in particular, the space of closed curves when $p_1 = p_2$ and $x_1 = x_2$. Now, in Ω we consider the action that measures the total curvature of curves, that is,

$$\mathcal{F} : \Omega \to \mathbb{R}, \qquad \mathcal{F}(\delta) = \int_\delta \kappa(s)\, ds,$$

where κ denotes the curvature function of curves in Ω. A natural problem is the study of the variational problem associated with this functional and specially to determine the curves which are extremals of this action.

Given $\delta \in \Omega$, the tangent space, $T_\delta\,\Omega$, of Ω at δ is identified with the space of vector fields along δ which vanish at the end points. If $W \in T_\delta\,\Omega$, then one can define a curve in Ω passing through δ in the direction of W, say for instance

$$\Gamma : (-\varepsilon, \varepsilon) \to \Omega, \qquad \Gamma(t) = \exp_{\delta(s)} tW(s)\,,$$

and, the first variation formula is defined by $D\mathcal{F}(\delta) : \mathbf{T}_\delta\,\Omega \to \mathbb{R}$

$$D\mathcal{F}(\delta)[W] = \frac{\partial}{\partial t}\left(\mathcal{F}(\Gamma(t))\right)\Big|_{t=0}.$$

It can be computed, using a standard argument which involves some integration by parts, and was set down in [2]

$$D\mathcal{F}(\delta)[W] = \int_\delta \langle \mathcal{E}(\delta), W\rangle ds + \mathcal{B}(\delta, W),$$

where $\mathcal{E}(\delta)$ and $\mathcal{B}(\delta, W)$ denote the so called Euler-Lagrange and boundary operators, respectively. The Frenet equations yield (see [2] for details)

$$\mathcal{E}(\delta) = \tau^2 N - \tau_s B - \tau\eta - R(N,T)T,$$

$$\mathcal{B}(\delta, W) = \sum_{i=1}^{m}\langle \nabla_T W(s_i), N(s_i^+) - N(s_i^-)\rangle + \sum_{i=1}^{m}\langle W(s_i), \nabla_T N(s_i^+) - \nabla_T N(s_i^-)\rangle,$$

where ∇ and R are, respectively, the Levi-Civita connection and the Riemannian curvature of the metric $\langle\,,\rangle$ on M. Moreover, T and N stand for the unit tangent and the unit normal (which is well defined if $n = 2$, while perhaps it is not defined in the inflections points $\delta(s_i)$, $1 \le i \le m$, when $n \ge 3$). Also B is the unit binormal and in this case $\tau \ge 0$ denotes the torsion and τ_s its derivative with respect to the arc length parameter. Finally, η belongs to the subbundle normal to the one spanned by $\{T, N, B\}(s)$ along $\delta(s)$, except at most at the inflection points. Now, the last formulas allow us to characterize the extremals for $\mathcal{F}$ in Ω. In fact, δ is a critical point of $\mathcal{F}$ in Ω if and only if the following conditions hold (see [2])

(1) N, B and τ are well defined at the inflection points of δ, and
(2) the following Euler-Lagrange equation is satisfied

$$R(N,T)T = \tau^2 N - \tau_s B - \tau\eta. \tag{4.1}$$

It seems clear that the degree of integrability we have, on the above stated field equation, strongly depends on the control we get on the curvature of the target space. Note that the influence of this space on the equation is exerted through the sectional curvature of the osculating plane of the trajectories and along them.

Let us briefly survey what is known on this equation and so on the critical points of the total curvature action. If M is the Euclidean plane, $\mathbb{R}^2$, then Eq. (4.1) becomes into an identity and this is so because, if fluctuations do not change the topology of curves, then the action $\mathcal{F}$ is constant on each homotopy class of curves in Ω. For example, if Ω consists of closed curves, the action $\mathcal{F}$ provides an integer multiple (the rotation index) of 2π. In the general case that value must be corrected by the angle between x_1 and x_2.

More generally, if M has dimension two, then (4.1) can be written as $K(\delta(s)) = 0$, where K stands for the Gaussian curvature of the surface. Therefore, extremal curves are made up of parabolic points. This case, including the stability of extremals, has been amply considered in [3].

On the other hand, the dynamics associated with $\mathcal{F}$ in a space of constant curvature c, say $M^n(c)$, for any dimension n, was completely solved in [2]. Let us briefly summarize it. First of all, note that in this case, the Euler-Lagrange equation (4.1) turns out to be

$$\tau^2 = c, \qquad \tau_s = 0, \qquad \tau\,\eta = 0,$$

and the third equation implies that the dynamics evolves in backgrounds with dimension $n \leq 3$. Moreover, the trajectories have constant torsion and $c \geq 0$. Next, it is clear that there are, up to topology, two possibilities to get extremals: the Euclidean plane, where the dynamics is trivial, and a round 3-sphere. This reduces the problem to determine those curves in a 3-sphere, say with radius one, whose torsion is, up to orientation, one. These curves were geometrically obtained in [2]: They are congruent to a horizontal lifting (also called a Legendrian curve), via the Hopf map, of a curve in the round two sphere with radius $1/2$.

Since the round 3-sphere is, essentially, the only constant curvature Riemannian space where the variational problem makes sense and, on the other hand, it is also a homogeneous 3-space, it seems natural to study the existence of extremals for the total curvature of curves in homogeneous 3-spaces M^3. It is known that the isometry group, $I(M^3)$, of M^3 is a Lie group whose dimension is 6, 4 or 3. From now on, we will say that it has rigidity of order 6, 4 or 3, respectively. In the first case, the space has constant curvature and so it is covered by either $\mathbb{S}^3$ (positive curvature), $\mathbb{R}^3$ (vanishing curvature) or $\mathbb{H}^3$ (negative curvature). So a natural breeding ground to study our variational problem is provided by the homogeneous 3-spaces in the second rigidity level, that is, when $\dim(I(M^3)) = 4$. This is certainly the widest and most interesting family of homogeneous 3-spaces, which sometimes are called rotationally symmetric homogeneous 3-spaces. Among its simply connected members, one can find, besides the Riemannian products $\mathbb{S}^2 \times \mathbb{R}$ and $\mathbb{H}^2 \times \mathbb{R}$, the Berger spheres, the Heisenberg group, Nil_3, and the universal covering of the special lineal group $\mathrm{SL}(2, \mathbb{R})$. Obviously, the family also includes quotient of these spaces by suitable isometry subgroups, namely,

- Lens spaces, $\mathrm{L}_n = \mathbb{S}^3/\mathbb{Z}_n$, $n \geq 2$, including the projective space, $\mathbb{RP}^3 = \mathrm{L}_2$, with the corresponding induced Berger metrics;
- Heisenberg bundles, including those over flat tori; and
- The projective special linear group $\mathrm{PSL}(2, \mathbb{R}) = \widetilde{\mathrm{PSL}}(2, \mathbb{R})/\mathbb{Z}_2$ and other quotients of $\widetilde{\mathrm{PSL}}(2, \mathbb{R})$.

Any homogeneous 3-space M^3 with rigidity of order four can be viewed as a bundle over a surface with constant curvature. To exploit this approach, we will follow the notation of [34]. This fibration provides a Riemannian submersion, $\mathrm{p} : M^3 \to \mathrm{B}^2(c)$, with geodesic fibers (vanishing first O'Neill invariant) over a constant curvature (say c) surface. In addition, the vertical flow is generated by a unit Killing vector field ξ which allows one to give the following expression for the second O'Neill invariant, A,

$$A_X\xi = \overline{\nabla}_X \xi = r\,(X \times \xi),$$

where X is a horizontal vector field and r, the bundle curvature [34], is a constant. Therefore, $\mathrm{p} : M^3 \to \mathrm{B}^2(c)$ is a Killing submersion in the sense of [39] and [73]. Let us denote by $\overline{\nabla}$ and ∇ the Levi-Civita connections on M^3 and $\mathrm{B}^2(c)$, respectively. Both constants, c (curvature of the base) and r (bundle curvature), classify the homogeneous 3-spaces up to isometries and topology. In other words, each pair of real numbers, (c, r), determines, up to topology, a congruence class $\mathrm{E}(c, r)$ of homogeneous 3-spaces whose isometry group has either dimension 4, if $c \neq 4r^2$, or dimension 6 (constant curvature), if $c = 4r^2$. From now on we will focus on homogeneous three spaces, $\mathrm{E}(c, r)$, whose isometry group has dimension 4 and consequently $c \neq 4r^2$. Furthermore, we will consider oriented homogeneous three spaces and, in particular, we will use orientation to define the cross product.

4.2 Lancret Helices in Homogeneous 3-Spaces

In a Riemannian space endowed with a unit Killing vector field ξ (also called an infinitesimal translation), curves making a constant angle with ξ are known as Lancret helices with axis ξ (see [6] for details on Lancret helices in the 3-sphere). As every homogeneous 3-space admits a unit Killing vector field, it seems natural to study the following (see [16]):

Problem 4.2.1 *Are there Lancret helices being extremal of the total curvature action in rotationally symmetric homogeneous spaces?*

This problem will be answered in the next section (see Proposition 4.3.2). To this end, it will be useful to characterize, geometrically, Lancret helices in homogeneous 3-spaces. Let $M^3 = \mathrm{E}(c, r)$ be a homogeneous 3-space, $\mathrm{p} : M^3 \to \mathrm{B}^2(c)$ the corresponding Riemannian submersion, and let ξ be the associated unit vertical Killing vector field. To study Lancret helices with axis ξ, it is useful to consider the so called equivariant surfaces. For any profile curve, $\beta(s)$, in $\mathrm{B}^2(c)$, choose a horizontal lifting, say $\bar{\beta}(s)$. Both curves have the same speed, so we will assume that both are parameterized by the arc-length. Now, consider the surface $S_\beta = \mathrm{p}^{-1}(\beta)$ in M^3, which can be nicely parameterized by horizontal liftings of β and ξ-orbits (from now on, simply orbits) via the following map

$$X(s, t) = \phi_t(\bar{\beta}(s)),$$

where $\{\phi_t : t \in \mathbb{R}\}$ is the one-parameter group generated by ξ. It is clear that S_β is an oriented flat surface, with the obvious induced orientation $X_s \times X_t$, which we will call the Hopf cylinder with profile curve $\beta(s)$. Now, it is easy to compute (see [39]) the second fundamental form of a Hopf cylinder, whose matrix in the above parameterization is given by

$$\begin{pmatrix} \kappa_g & -r \\ -r & 0 \end{pmatrix}$$

where κ_g stands for the geodesic curvature of the profile curve in the base $\mathtt{B}^2(c)$.

Let ∇' be the Levi-Civita connection of the induced metric on S_β and let $\gamma(s)$ be a unit speed geodesic in a certain Hopf cylinder, S_β. Then $\nabla'_{\gamma'}\gamma'(s) = 0$ and so we have

$$\frac{d}{ds}\langle\gamma'(s), \xi(\gamma(s))\rangle = \langle\overline{\nabla}_{\gamma'}\gamma'(s), \xi(\gamma(s))\rangle + r\langle\gamma'(s), \gamma'(s) \times \xi(\gamma(s))\rangle = 0,$$

which shows that the angle that $\gamma(s)$ makes with ξ is constant. In other words, *the geodesics of the Hopf cylinders are Lancret helices.*

To show the converse, let $\gamma(s)$ be a unit speed Lancret helix, so there exists $\varphi \in \mathbb{R}$, which we may suppose different from zero (otherwise the result is trivial) such that $\langle\gamma', \xi\rangle(s) = \cos\varphi$. Now, define the curve $\beta(s) = \mathtt{p}(\gamma(s))$ in $\mathtt{B}^2(c)$ and use it as a profile curve to get the Hopf cylinder $S_\beta = \mathtt{p}^{-1}(\beta)$. It is clear that the original Lancret helix lies in S_β, and, moreover, we have

$$0 = \frac{d}{ds}\langle\gamma', \xi\rangle(s) = \langle\overline{\nabla}_{\gamma'}\gamma', \xi\rangle(s) + r\langle\gamma', \gamma' \times \xi\rangle(s) = \langle\nabla'_{\gamma'}\gamma', \xi\rangle(s),$$

which proves that $\nabla'_{\gamma'}\gamma'(s) = 0$, and so $\gamma(s)$ is a geodesic in $S_\beta = \mathtt{p}^{-1}(\beta)$. Therefore we have the following

Proposition 4.2.2 *The Lancret helices in any homogeneous 3-space with rigidity of order four are just the geodesics of Hopf cylinders.*

Next, we wish to compute the Frenet apparatus, $\{T_\gamma, N_\gamma, B, \kappa, \tau\}(s)$ of a Lancret helix $\gamma(s)$ in M^3. Let $\beta(u)$ be the unit speed profile curve such that $\gamma(s)$ is a geodesic in $S_\beta = \mathtt{p}^{-1}(\beta)$. Furthermore, $\{T = \beta', N = JT, \kappa_g\}$ will be the Frenet apparatus of $\beta(u)$ in $\mathtt{B}^2(c)$. We also have the above standard parameterization $X(u,t) = \phi_t(\bar{\beta}(u))$ on the Hopf cylinder, so that $\overline{T}(u,t) = X_u(u,t) = d\phi_t(\bar{\beta}'(u))$ is a horizontal lifting of T along the Hopf cylinder. In this context, since $\gamma(s)$ is a geodesic in the flat surface $S_\beta = \mathtt{p}^{-1}(\beta)$, it must be the image under X of a certain straight line in the (u,t)-plane. Therefore, there exists $\varphi \in (0, 2\pi)$ such that $\gamma(s) = X((\sin\varphi)\,s, (\cos\varphi)\,s)$. Then we have $T_\gamma = \sin\varphi\,\overline{T} + \cos\varphi\,\xi$. We also know that $N_\gamma = \overline{T} \times \xi$, because it must agree with the unit normal to the Hopf cylinder, and therefore $B = \cos\varphi\,\overline{T} - \sin\varphi\,\xi$. To compute the curvature function, $\kappa(s)$, we use that it should be the normal curvature of γ in S_β and then

$$\kappa = \begin{pmatrix} \sin\varphi & \cos\varphi \end{pmatrix} \begin{pmatrix} \kappa_g & -r \\ -r & 0 \end{pmatrix} \begin{pmatrix} \sin\varphi \\ \cos\varphi \end{pmatrix} = \frac{\kappa_g - 2rm}{m^2 + 1}, \qquad (4.2)$$

where $m = \cot\varphi$. Finally, a direct computation allows us to obtain

$$\overline{\nabla}_{T_\gamma} B = [\kappa_g \cos\varphi \sin\varphi - r(\cos^2\varphi - \sin^2\varphi)]\, N_\gamma,$$

which gives

$$\tau = \frac{1}{2}\kappa_g \sin 2\varphi - r\cos 2\varphi = \frac{\kappa_g\, m - r(m^2 - 1)}{m^2 + 1}. \tag{4.3}$$

It is worth pointing out that, from (4.2) and (4.3), the curvature and torsion functions of Lancret helices in homogeneous 3-spaces are nicely related by

$$\tau = (\cot\varphi)\,\kappa + r. \tag{4.4}$$

Moreover, (4.2) and (4.3) allow us to compute the curvature, κ_g, of the profile curve and the slope, $m = \cot\varphi$, of a Lancret helix in terms of both its curvature function, κ, and its torsion τ. In fact, we get

$$\kappa_g = \kappa + \frac{\tau^2 - r^2}{\kappa}, \qquad m = \frac{\tau - r}{\kappa}.$$

As a consequence, we have the following

Proposition 4.2.3 *A Lancret helix in a homogeneous 3-space $M^3 = \mathrm{E}(c, r)$ is unique and equivalently determined by either*

(i) Its curvature and torsion functions (κ, τ); or
(ii) The curvature, κ_g, of the profile curve in the space of orbits and the slope of the helix, $m = \cot\varphi$, as a geodesic in the corresponding equivariant surface.

The classical theorem of Lancret ensures that formula (4.4) with $r = 0$ provides a simple characterization of Lancret helices in the Euclidean space. This classical theorem of Lancret was discovered to be true in the 3-sphere (see [6]), where it was also observed that the only Lancret helices in the hyperbolic 3-space are those with curvature an torsion being both constant (in some sense they are like circular helices). Thus, it seems natural to ask for a Lancret-type theorem in homogeneous 3-spaces with rigidity of order four. In other words

Problem 4.2.4 *Does equation* (4.4) *characterize Lancret helices in such homogeneous 3-spaces?*

However, a rotationally symmetric homogeneous space does not have enough isometries to guarantee the converse and so the theorem of Lancret. More precisely, we are going to show the existence of a curve, in a homogeneous 3-space with rigidity of order four, with the same curvature and the same torsion of a Lancret helix, but they are not congruent.

Proposition 4.2.5 *Let $M^3 = \mathrm{E}(c, r)$ be an homogeneous 3-space with $c \neq 4r^2$, and let $\gamma(s)$ be a Lancret helix in M^3 with curvature $\kappa(s)$ and torsion $\tau(s)$. Then there exists a curve $\delta(s)$ in M^3 with curvature function $\kappa_\delta = \kappa$ and torsion $\tau_\delta = \tau$ which is not congruent to any Lancret curve in M^3.*

Proof The Frenet frame $\{T(s), N(s), B(s)\}$ of a curve parameterized by the arc-length, in M^3, with curvature $\kappa(s)$ and torsion $\tau(s)$ comes as a solution of the Frenet equations

$$\overline{\nabla}_s \begin{pmatrix} T(s) \\ N(s) \\ B(s) \end{pmatrix} = \begin{pmatrix} 0 & \kappa(s) & 0 \\ -\kappa(s) & 0 & \tau(s) \\ 0 & -\tau(s) & 0 \end{pmatrix} \begin{pmatrix} T(s) \\ N(s) \\ B(s) \end{pmatrix}$$

which is a first order linear system of differential equations. For any $p \in M^3$ and an orthonormal frame $\{\vec{t}, \vec{n}, \vec{b}\}$ in the tangent space $T_p(M^3)$, there exists a unique solution, $\{T_\delta(s), N_\delta(s), B_\delta(s)\}$, of the above system and then a unique arc-length parameterized curve, $\delta(s)$, such that $\delta(0) = p$ and $\{T_\delta(0), N_\delta(0), B_\delta(0)\} = \{\vec{t}, \vec{n}, \vec{b}\}$. Now, suppose that $\gamma(s)$ is a Lancret helix in M^3 with curvature $\kappa(s)$ and torsion $\tau(s)$ and let $m = \cot\varphi$ be its slope as a geodesic in a suitable Hopf cylinder. Now, by choosing an initial condition such that the angle between $\vec{t}$ and $\xi(p)$ is different from φ, then the corresponding solution $\delta(s)$ is not congruent to any Lancret helix. In fact, note that the slope of a Lancret helix is completely determined by

$$m = \cot\varphi = \frac{\tau - r}{\kappa}.$$

□

4.3 Extremals of the Total Curvature in Homogeneous 3-Spaces

The Riemannian curvature operator of a homogeneous 3-space $M^3 = \mathrm{E}(c, r)$ was computed in [34] to be

$$\begin{aligned} R(X, Y)Z = (c - 3r^2)(\langle Y, Z\rangle X - \langle X, Z\rangle Y) \\ + (-c + 4r^2)[\langle Y, \xi\rangle\langle Z, \xi\rangle X - \langle X, \xi\rangle\langle Z, \xi\rangle Y \\ + \langle X, \xi\rangle\langle Y, Z\rangle\xi - \langle Y, \xi\rangle\langle X, Z\rangle\xi]. \end{aligned}$$

In particular, along a curve $\gamma(s)$ with Frenet frame $\{T, N, B\}(s)$, we have

$$R(N, T)T = (c - 3r^2)N + (-c + 4r^2)[\langle T, \xi\rangle^2 N - \langle T, \xi\rangle\langle N, \xi\rangle T + \langle N, \xi\rangle\xi],$$

which can be combined with (4.1) to obtain the following equation whose solutions provide the extremals of the total curvature action on M^3 in $\mathrm{E}(c, r)$

$$\tau^2 = r^2 + (c - 4r^2)\langle B, \xi\rangle^2, \qquad \tau_s = (c - 4r^2)\langle N, \xi\rangle\langle B, \xi\rangle.$$

We can manipulate these equations to see that those extremals are just the solutions of the following field equations

$$\tau^2 = r^2 + (c - 4r^2)\langle B, \xi\rangle^2, \qquad (r - 2\tau)\langle N, \xi\rangle\langle B, \xi\rangle = 0. \tag{4.5}$$

The following result reduces the integration of the above equations to three fundamental cases.

Lemma 4.3.1 *Let $\gamma(s)$ be an extremal of the total curvature functional in* $\mathrm{E}(c, r)$, *then either*

(E1) *ξ lies in its rectifying plane along $\gamma(s)$, or*
(E2) *ξ lies in its osculating plane along $\gamma(s)$.*

Proof According to the second equation in (4.5), we only need to study the case $r - 2\tau = 0$. Then the first equation of (4.5) shows that

$$\langle B, \xi\rangle^2 = -\frac{3\tau^2}{c - 4r^2},$$

and so $\langle B, \xi\rangle$ is a constant. However $\frac{d}{ds}\langle B, \xi\rangle = (r - \tau)\langle N, \xi\rangle$ and consequently the extremals in this class have either horizontal normal or torsion $\tau = r$. However in the last case $\tau = 0$ and so the binormal is horizontal. □

4.3.1 Extremals with Horizontal Normal

First of all, observe that a curve in $\mathrm{E}(c, r)$ has horizontal normal if and only if it is a Lancret helix. That follows from

$$\frac{d}{ds}\langle T, \xi\rangle = \kappa\langle N, \xi\rangle + r\langle T, T \times \xi\rangle = \kappa\langle N, \xi\rangle,$$

where geodesics can be considered trivial Lancret helices, as well as trivial extremals of the total curvature. In fact, they are minima, because the curvature function of a curve in dimension three does not change sign.

Consequently, the class of extremals with horizontal normal is a subclass of the family of Lancret helices. To determine those Lancret helices providing extremals, we only need to check the first equation of (4.5) because the second equation automatically holds for Lancret helices. Now, for a Lancret helix we see that

$\langle B, \xi \rangle = \sin \varphi = \frac{1}{\sqrt{m^2+1}}$ is constant and therefore the curvature and torsion of extremal Lancret helices are both constant functions. Therefore, we obtain the following characterization of extremal Lancret helices.

Proposition 4.3.2 *A Lancret helix in* $\mathrm{E}(c, r)$ *is an extremal of the total curvature if and only if it has constant torsion satisfying that*

$$\tau^2 = \frac{c + r^2(m^2 - 3)}{m^2 + 1}. \tag{4.6}$$

Remark 4.3.3 Since the only obstruction to the existence of extremal Lancret helices in $\mathrm{E}(c, r)$ is provided, according to the above proposition, by the positivity of $c + r^2(m^2 - 3)$, we have the following

Corollary 4.3.4 *Every homogeneous* 3*-space* $\mathrm{E}(c, r)$*, except* $\mathbb{H}^2 \times \mathbb{R}$*, admits a real one-parameter class of Lancret helices which are extremal of the total curvature action.*

Proof Homogeneous 3-spaces which are Riemannian products correspond to $r = 0$. Then for $\mathbb{H}^2 \times \mathbb{R}$, as $c < 0$, the above proposition shows the non-existence of extremal Lancret helices. Otherwise, we have the real one-parameter class of extremal Lancret helices with slope satisfying

$$m^2 > \frac{3r^2 - c}{r^2}.$$

□

Now, helices in this class can be geometrically obtained as follows. The constancy of both torsion and curvature imply that they are geodesics in Hopf cylinders built over curves with constant geodesic curvature (circles) in $\mathrm{B}^2(c)$. According to (4.3) the torsion of any Lancret helix can be expressed in terms of both the geodesic curvature, κ_g, of the profile curve in $\mathrm{B}^2(c)$ and the slope, $m = \cot \varphi$, of the helix as a geodesic in the corresponding Hopf cylinder. Therefore, for any m satisfying the above inequality, just choose the geodesic with slope $m = \cot \varphi$ in the Hopf cylinder S_{β_m}, where β_m is the circle in $\mathrm{B}^2(c)$ with geodesic curvature

$$\kappa_g = \frac{\sqrt{(m^2 + 1)(c + r^2(m^2 - 3))} + r(m^2 - 1)}{m}, \tag{4.7}$$

to get a Lancret helix which is an extremal of the total curvature in $\mathrm{E}(c, r)$. □

Example 4.3.5 In Riemannian product homogeneous 3-spaces, we have $r = 0$ and then we observe that while $\mathbb{H}^2(c) \times \mathbb{R}$ has no extremal Lancret helices, in $\mathbb{S}^2(c) \times \mathbb{R}$ we can find a one-parameter class of Lancret helices which are extremal of the total curvature energy. In fact, for any $m \in \mathbb{R} - \{0\}$, we choose in $\mathbb{S}^2(c)$ the circle β_m with geodesic curvature

$$\kappa_g = \frac{\sqrt{c(m^2+1)}}{m},$$

and then take γ_m as the geodesic with slope m in the Hopf cylinder shaped on β_m. Then γ_m is an extremal of the total curvature in $\mathbb{S}^2(c) \times \mathbb{R}$.

Example 4.3.6 In nilmanifold geometries, which are associated to the Heisenberg group, we have $c = 0$. Therefore, for any $\varphi \in (0, \pi/6)$ there exists a Lancret helix which is an extremal of the total curvature action. These extremals appear as geodesic in Hopf cylinders built over circles in the Euclidean plane with curvature

$$\kappa_g = \frac{r}{m}\left(\sqrt{m^4 - 2m^2 - 3} + m^2 - 1\right), \qquad m = \cot\varphi.$$

4.3.2 *Extremals with Horizontal Binormal*

In a three dimensional oriented Riemannian space one can define the cross product. Actually we are considering $\mathrm{E}(c, r)$ as an oriented space and have used the cross product to define the unit binormal of a curve, $\gamma(s)$, as $B(s) = T(s) \times N(s)$. In particular, the osculating planes along $\gamma(s)$ admit the obvious induced orientation $\{T(s), N(s)\}$.

In this section we focus on the study of extremals with horizontal binormal, $\langle B, \xi \rangle = 0$, and observe that any curve in $\mathrm{E}(c, r)$ satisfying this condition is automatically an extremal of the total curvature functional. Indeed, just note that

$$\frac{d}{ds}\langle B, \xi \rangle = (r - \tau)\langle N, \xi \rangle = 0,$$

and so $\tau = r$. Otherwise, the curve is an orbit. Thus the first equation of (4.5) holds. Consequently the whole family of curves with horizontal binormal in $\mathrm{E}(c, r)$ constitutes a class of extremals of the total curvature energy. Now, our immediate purpose is to understand this class of extremals.

Let $\gamma(s)$ be a unit speed curve with horizontal binormal. Then ξ lies in the osculating plane along $\gamma(s)$ and it allows us to define a unit horizontal vector field, η, along the curve, which also lies in the osculating plane, and satisfies that $\eta \times \xi = B$. We call $\{B, \eta, \xi\}$ a positive oriented orthonormal frame along $\gamma(s)$ which is adapted to the horizontal-vertical decomposition. Therefore, $\{T, N\}$ and $\{\eta, \xi\}$ provide the same orientation in the osculating plane along $\gamma(s)$. Let $\psi(s)$, $0 < \psi(s) < 2\pi$, be the angle between $T(s) = \gamma'(s)$ and ξ. Then it is a locally well defined differentiable function satisfying that

$$\begin{pmatrix} \eta(s) \\ \xi(s) \end{pmatrix} = \begin{pmatrix} \sin\psi(s) & -\cos\psi(s) \\ \cos\psi(s) & \sin\psi(s) \end{pmatrix} \begin{pmatrix} T(s) \\ N(s) \end{pmatrix} \tag{4.8}$$

Next we have the following characterization of the horizontality of the binormal:

Proposition 4.3.7 *Given a curve $\gamma(s)$ in $\mathrm{E}(c, r)$, the following statements are equivalent:*

(a) It has horizontal binormal;
(b) Its curvature function $\kappa(s)$, satisfies that $\kappa(s) + \psi'(s) = 0$;
(c) It is an extremal of the total curvature with $\tau = r$.

Proof From (4.8) we have $\langle T(s), \xi\rangle = \cos\psi(s)$ and then $\kappa\langle N(s), \xi\rangle + \psi'(s)\sin\psi(s) = 0$. Moreover since $\xi = \cos\psi\, T + \sin\psi\, N$, we obtain $(\kappa + \psi')\sin\psi = 0$.
Conversely, if $\kappa(s) + \psi'(s) = 0$, then $\kappa(\langle N(s), \xi\rangle - \sin\psi) = 0$, and $\langle N(s), \xi\rangle = \sin\psi(s)$, which implies that $\langle B, \xi\rangle = 0$. This shows the equivalence between (a) and (b).
The equivalence between (a) and (c) follows from (4.5). □

Remark 4.3.8

(1) It should be observed that, as a consequence of the above proposition, curves in $\mathrm{E}(c, r)$ with curvature function satisfying $\kappa(s) = -\psi'(s)$, automatically have constant torsion $\tau = r$.
(2) Note also the existence of curves in $\mathrm{E}(c, r)$ with $\tau = r$ whose binormal is not horizontal. For example, the horizontal liftings of any curve in $\mathrm{B}^2(c)$.

To describe this class of curves from a geometric point of view and how they can be obtained, we proceed as follows. Let $\gamma(s)$ be a curve in $\mathrm{E}(c, r)$ whose binormal is horizontal and let $\gamma_1(s) = \pi(\gamma(s))$ be its projection over $\mathrm{B}^2(c)$, which, in general, need not be a unit speed curve. However, $B_1 = d\pi(B)$ defines a unit normal vector field along γ_1 and so $JB_1 = \gamma_1'/|\gamma_1'|$ is the unit tangent of that projection, where J denotes the usual complex structure on $\mathrm{B}^2(c)$. Let JB be the unit horizontal vector field along $\gamma(s)$ satisfying that $d\pi(JB) = JB_1$. Therefore, $\{B, JB, \xi\}$ provides a positive oriented orthonormal frame along $\gamma(s)$ which is adapted to the horizontal-vertical decomposition. In this framework, the unit tangent and the unit principal normal of $\gamma(s)$ in $\mathrm{E}(c, r)$ are given by

$$T(s) = \cos\psi\,\xi + \sin\psi\, JB, \qquad N(s) = \sin\psi\,\xi - \cos\psi\, JB.$$

Now, we compute the curvature function, $\kappa_g(s)$, of $\gamma_1(s)$ in $\mathrm{B}^2(c)$, which will determine, up to motions in $\mathrm{B}^2(c)$, the profile curve in whose Hopf cylinder $\gamma(s)$ lies. Then we have

$$\overline{\nabla}_T T = -\psi' \sin\psi\, \xi + r \cos\psi \sin\psi\, B + \psi' \cos\psi\, JB + \sin\psi\, \overline{\nabla}_T JB,$$

and

$$\overline{\nabla}_T JB = (r \cos\psi + \kappa_g \sin\psi)\, B.$$

Now, we use the first Frenet equation of $\gamma(s)$ in $\mathrm{E}(c, r)$, with $\kappa(s) = -\psi'(s)$, to see that

$$\kappa_g(s) = -2r \cot\psi(s).$$

At this point, the way to construct the class of extremals with horizontal binormal works as follows:

(1) First, we choose an arbitrary positive function $h(s)$, which will play the role of the curvature function of the extremal, and define the function

$$\psi(s) = -\int_0^s h(v)\, dv.$$

(2) Next, we take the curve $\beta(s)$ in $\mathrm{B}^2(c)$ (it is unique up to rigid motions) which is determined by its arc-length function $u(s)$ and its curvature function $\kappa(s)$ given, respectively, by

$$u(s) = \int_0^s \sin\psi(v)\, dv, \qquad \kappa_g(s) = -2r \cot\psi(s).$$

(3) Then, we use $\beta(s)$ as a profile curve to construct its Hopf cylinder S_β in $\mathrm{E}(c, r)$ and then choose in this flat surface the curve, $\gamma(s)$, with slope function (measured with respect to ξ) $\psi(s)$. We conclude that $\gamma(s)$ is an extremal of the total curvature functional on $\mathrm{E}(c, r)$ with horizontal binormal (consequently $\tau = r$) and curvature function $h(s)$. Moreover, all of extremals with horizontal binormal are obtained in this way.

It is well known that the curvature function of a plane curve is just the variation of an angle function measured with respect to a fixed direction and this is the key point to recover the curve, through quadratures, from its curvature function. So the solving natural equation works for plane curves. Now, the curvature function of a curve in $\mathrm{E}(c, r)$ with horizontal binormal is the variation of the angle that it makes with the Reeb vector field, ξ, and this is the main point to solve the so called solving natural equations problem for this kind of curve in $\mathrm{E}(c, r)$. For a better understanding of this problem, recall that for any curve $\beta(s)$ in $\mathrm{B}^2(c)$, we can choose a horizontal lifting $\bar{\beta}(s)$ in $\mathrm{E}(c,r)$ and then parameterize the corresponding Hopf cylinder S_β by

$$X : \mathbb{R}^2 \to S_\beta \subset \mathrm{E}(c,r), \qquad X(u,t) = \phi_t(\bar{\beta}(u)),$$

where $\{\phi_t, : t \in \mathbb{R}\}$ is the one-parameter subgroup associated with ξ. Now, curves in $\mathbb{R}^2$ can be nicely obtained, using quadratures, from their curvature functions. In particular, the unit speed curve with curvature function $h(s)$ is given by

$$\alpha(s) = (u(s), t(s)) = \left(\int_0^s \sin\psi(v)\, dv, \int_0^s \cos\psi(v)\, dv \right),$$

where $\psi(s) = -\int_0^s h(v)\, dv$ is the angle that $\alpha(s)$ makes with the t-axis. Since X is an isometry, $\psi(s)$ is the angle that the curve

$$\gamma(s) = X(\alpha(s)) = \phi_{\int_0^s \cos\psi(v)\, dv} \left(\bar{\beta} \left(\int_0^s \sin\psi(v)\, dv \right) \right),$$

makes with the orbits. The curvature $\kappa(s)$ of $\gamma(s)$ in $\mathrm{E}(c,r)$ does not necessarily satisfy $\kappa(s) + \psi'(s) = 0$. However, it holds if and only if the curve $\beta(s)$ is chosen to have arc-length and curvature functions given, respectively, by

$$u(s) = \int_0^s \sin\psi(v)\, dv, \qquad \kappa_g(s) = -2r \cot\psi(s).$$

Remark 4.3.9 It should be noted that this argument can be viewed as the solution of the so called solving natural equations for curves with horizontal binormal in $\mathrm{E}(c,r)$ problem. In fact, we obtain explicitly the curve from its curvature function through quadratures. Certainly, it is a surprising result keeping in mind that the spaces $\mathrm{E}(c,r)$ have not the highest rigidity. Moreover, the next theorem measures the size of this class of curves.

Even more, we can exploit the above argument to exhibit a one-to-one correspondence between the class of convex plane curves and the family of extremals with horizontal binormal in $\mathrm{E}(c,r)$. It can be obtained from the following result

Theorem 4.3.10 *For every convex curve α in the Euclidean plane, there exists a curve γ in* $\mathrm{E}(c,r)$ *with the same curvature function as α and torsion $\tau = r$, which is an extremal, with horizontal binormal, of the total curvature action on* $\mathrm{E}(c,r)$.

Proof Let $\alpha(s) = (u(s), t(s))$ be any convex curve in the plane and denote by $\kappa(s) > 0$ its curvature function. Now, consider in $\mathrm{B}^2(c)$ a curve, $\beta(s)$ (unique up to motions) whose curvature function is $\kappa_g(s) = -2r \cot\psi(s)$ where

$$\psi(s) = -\int_0^s \kappa(v)\, dv.$$

Then consider the map $X : \mathbb{R}^2 \to S_\beta \subset \mathrm{E}(c, r)$ defined by

$$X(u, t) = \phi_t(\bar{\beta}(s),$$

where $\bar{\beta}$ is a horizontal lifting of β. Finally the curve $\gamma(s) = X(\alpha(s))$ has curvature function $\kappa(s)$, not only in S_β but also in $\mathrm{E}(c, r)$, its torsion is $\tau = r$ and it is an extremal of the total curvature in $\mathrm{E}(c, r)$ which certainly has horizontal binormal.

□

A relevant special case appears when $r = 0$. In this case the homogeneous 3-space is a Riemannian product $\mathrm{E}(c, 0) = \mathrm{B}^2(c) \times \mathbb{R}$, where (up to topology) $\mathrm{B}^2(c)$ is $\mathbb{S}^2(c)$ if $c > 0$ while it is $\mathbb{H}^2(c)$ when $c < 0$. Then, as a consequence of the above argument we have the following characterization for extremals with horizontal binormal

Corollary 4.3.11 *Every curve, $\gamma(s)$, with horizontal binormal in $\mathrm{B}^2(c) \times \mathbb{R}$ lies in a Hopf cylinder built on a geodesic of $\mathrm{B}^2(c)$. More precisely these curves can be explicitly parameterized as follows*

$$\gamma(s) = \left(\beta(u(s)), \int_0^s \cos\psi(v)\,dv\right),$$

where $\beta(u)$ is a unit speed geodesic in $\mathrm{B}^2(c)$ with arc-length function given by

$$u(s) = \int_0^s \sin\psi(v)\,dv \qquad \textit{with} \qquad \psi(s) = -\int_0^s \kappa(v)\,dv,$$

and where $\kappa(s)$ is an arbitrary function making the role of curvature function of $\gamma(s)$.

Observe that if β is a geodesic in $\mathrm{B}^2(c)$, then S_β is a totally geodesic surface in $\mathrm{E}(c, 0) = \mathrm{B}^2(c) \times \mathbb{R}$. On the other hand, extremals with horizontal binormal have torsion zero. Therefore, the first claim of the above corollary can be viewed as a codimension reduction result, which is the classical behavior in spaces with constant curvature. In contrast, extremals in $\mathbb{S}^2(c) \times \mathbb{R}$ with horizontal normal do not lie in any totally geodesic surface (see Example 4.3.5).

4.4 Some Examples and Applications

In this section, we will show a few examples of curves which are extremal for the total curvature energy in $\mathrm{E}(c, r)$. They attempt to clarify the results which have been obtained. We will not exhaust all possible cases, rather we will restrict ourselves to the two simplest ones, Riemannian products and nilmanifolds, where examples can be given explicitly and pictures can be exhibited (see [16]).

4.4.1 Extremals in Riemannian Products

Remember that, as we have shown previously, the whole class of extremals in $\mathbb{H}^2(c) \times \mathbb{R}$ is made up of curves with horizontal binormal (see Proposition 4.3.7). On the contrary, in $\mathbb{S}^2(c) \times \mathbb{R}$, besides the class of curves with horizontal binormal (see Corollary 4.3.4), another one-parameter class of extremals with horizontal unit normal also appears (see Example 4.5). This additional class is formed by Lancret helices having both constant curvature and torsion given, respectively, by

$$\tau = \sqrt{\frac{c}{m^2+1}}, \qquad \kappa = \frac{1}{m}\sqrt{\frac{c}{m^2+1}}.$$

We will use Proposition 4.3.7 to provide some explicit examples of curves with horizontal binormal, and so being extremal for the total curvature, in $\mathrm{B}^2(c) \times \mathbb{R}$ with either $c = 1$ or $c = -1$.

Let us write $\mathbb{S}^2 = \{(x, y, z) \in \mathbb{R}^3 : x^2 + y^2 + z^2 = 1\}$ and $\mathbb{H}^2 = \{(x, y, z) \in \mathbb{L}^3 : x^2 + y^2 - z^2 = -1,\ z > 0\}$ and consider, without loss of generality, the geodesics $\beta(u) = (\cos u, 0, \sin u) \subset \mathbb{S}^2$ and $\beta(u) = (\cosh u, 0, \sinh u) \subset \mathbb{H}^2$. Now, the algorithm to construct extremals with horizontal binormal in $\mathrm{B}^2 \times \mathbb{R}$, which is implicitly contained in Proposition 4.3.7, consists in building curves in the surface $\beta \times \mathbb{R}$ starting from a prescribed function that makes the role of curvature function. Applying this, one has, for instance

(1) **Circles viewed as extremals.** Consider a constant function $\kappa(s) = -a$, then we have $\psi(s) = as$ and therefore

$$\gamma(s) = (\cos(-\frac{1}{a}\cos(as)), 0, \sin(-\frac{1}{a}\cos(as)), \frac{1}{a}\sin(as))$$

is an extremal of the total curvature energy of curves in $\mathbb{S}^2 \times \mathbb{R}$, while

$$\gamma(s) = (\sinh(-\frac{1}{a}\cos(as)), 0, \cosh(-\frac{1}{a}\cos(as)), \frac{1}{a}\sin(as))$$

is an extremal in $\mathbb{H}^2 \times \mathbb{R}$.

(2) **Clothoids regarded as extremals.** Pick up the function $\kappa(s) = -s$, that represents the curvature function of a clothoid or Cornu spiral. We can solve the natural equations to obtain the clothoid in terms of the Fresnel integrals

$$\alpha(s) = (u(s), t(s)) = \left(\int_0^s \sin(v^2/2)\,dv, \int_0^s \cos(v^2/2)\,dv\right).$$

Since $\psi(s) = \frac{s^2}{2}$, we have that

$$\gamma(s) = \left(\cos\left(\int_0^s \sin(v^2/2)\,dv\right), 0, \sin\left(\int_0^s \sin(v^2/2)\,dv\right), \int_0^s \cos(v^2/2)\,dv\right),$$

is an extremal of the total curvature in $\mathbb{S}^2 \times \mathbb{R} \subset \mathbb{R}^4$, while

$$\gamma(s) = \left(\sinh\left(\int_0^s \sin(v^2/2)\,dv\right), 0, \cosh\left(\int_0^s \sin(v^2/2)\,dv\right), \int_0^s \cos(v^2/2)\,dv\right),$$

is an extremal of the total curvature in $\mathbb{H}^2 \times \mathbb{R} \subset \mathbb{L}^3 \times \mathbb{R}$.

(3) Catenaries as extremals. Now, take the function $\kappa(s) = \frac{1}{1+s^2}$. The unit speed curve having this curvature function is the catenary

$$\alpha(s) = (u(s), t(s)) = \left(\sqrt{1+s^2}, \log(s + \sqrt{1+s^2}\right).$$

In this case, $\psi(s) = \arctan s$, and, consequently, we obtain that the following curves

$$\gamma(s) = \left(\cos\sqrt{1+s^2}, 0, \sin\sqrt{1+s^2}, \log(s+\sqrt{1+s^2}\right),$$

$$\gamma(s) = \left(\sinh\sqrt{1+s^2}, 0, \cosh\sqrt{1+s^2}, \log(s+\sqrt{1+s^2}\right).$$

are extremal for the total curvature in $\mathbb{S}^2 \times \mathbb{R}$ and $\mathbb{H}^2 \times \mathbb{R}$, respectively.

4.4.2 Extremals in Nilmanifolds

Now, we wish to add a few examples describing extremals in nilmanifolds geometries which are associated with the Heisenberg group. In dimension three they are one of the eight canonical Thurston geometries. According to the usual terminology, they are identified with $\mathrm{E}(0, r)$ and, consequently, appear as Riemannian submersions over a Euclidean plane.

The 3-dimensional Heisenberg group, N_r^3, is nothing but $\mathbb{R}^3$ endowed with the metric $ds^2 = dx^2 + dy^2 + \theta^2$, where θ is the one-form defined by

$$\theta = dz + r(y\,dx - x\,dy).$$

It is known that $\xi = \partial_z$ is an infinitesimal translation, which is, precisely, the Reeb vector field when (N_r^3, θ) is viewed as a contact manifold. The one-parameter subgroup associated with ξ is

$$\{\phi_t : \mathrm{N}_r^3 \to \mathrm{N}_r^3 : t \in \mathbb{R}\}, \qquad \phi_t(x, y, z) = (x, y, z + t).$$

Every orbit is principal and all of them are geodesics of N_r^3. The space of orbits is the Euclidean plane and the projection $\pi : \mathrm{N}_r^3 \to \mathbb{R}^2$ is given by $\pi(x, y, z) = (x, y)$.

We recall again that we have two classes of extremals for the total curvature action on 3-dimensional homogeneous 3-spaces and that, in addition, we have provided algorithms and methods to construct extremals within both families. In particular, they can be described as curves lying in Hopf cylinders. This key fact will be used to give explicit extremals of the total curvature in the Heisenberg group.

Hopf cylinders shaped on a plane curve $\beta(s)$, S_β, are flat surfaces which can be obtained by specifying a Legendrian curve. So, choose a profile curve, $\beta(s) = (x(s), y(s))$, in the space of orbits, and assume that it is parameterized by the arc-length. Their horizontal liftings (Legendrian curves) $\bar{\beta}(s)$ project down over $\beta(s)$, and then $\bar{\beta}(s) = (x(s), y(s), z(s))$, where horizontality implies that the third coordinate must satisfy $\theta(\bar{\beta}'(s)) = 0$. Hence, horizontal liftings turn out to be determined by the solutions of the differential equation

$$\frac{dz}{ds} = r\left(x\frac{dy}{ds} - y\frac{dx}{ds}\right). \tag{4.9}$$

Thus, a horizontal lifting of the profile curve (which is obviously parameterized by the arc-length too) provides the following natural parameterisation of the equivariant surface S_β

$$X(s, t) = \phi_t(\bar{\beta}(s)) = (x(s), y(s), z(s) + t).$$

Now, the first family of extremals of the total curvature action we alluded to before (critical curves with horizontal unit normal) are Lancret helices. Then, we apply the corresponding algorithm, given in Sect. 4.3.1, to obtain a real one-parameter family of extremals in the 3-dimensional Heisenberg group, namely we have

Corollary 4.4.1 *For any $\varphi \in (0, \pi/6)$, there exists a Lancret helix in N_r^3, with slope $m = \cot\varphi$ in a suitable Hopf cylinder, which is an extremal of the total curvature.*

Lancret helices in N_r^3 are curves, $\gamma(s)$, forming a constant angle with ξ and they coincide precisely with the set of geodesics of the Hopf cylinders of N_r^3 (Propositon 4.2.2). In particular, those Lancret helices which provide extremals for the total curvature must be geodesics of Hopf cylinders shaped on circles of the Euclidean plane. So, let $\beta(s) = \left(R\cos\frac{s}{R}, R\sin\frac{s}{R}\right)$ be the unit speed circle with radius R and centered at the origin. In this case, it is very easy to solve the Eq. (4.9) to see that horizontal liftings of the above circles are just circular helices when they are viewed in the Euclidean 3-space, namely

$$\bar{\beta}(s) = \left(R\cos\frac{s}{R}, R\sin\frac{s}{R}, rRs + c\right), \qquad \forall c \in \mathbb{R}.$$

Remark 4.4.2 Notice that Euclidean horizontal liftings of $\beta(s)$ are just circles, with radius R and centers on the z-axis, located in planes parallel to $z = 0$.

The corresponding Hopf cylinders can be parameterized by

$$X : \mathbb{R}^2 \to \mathrm{N}^3, \qquad X(s,t) = \left(R\cos\frac{s}{R}, R\sin\frac{s}{R}, rRs + t\right).$$

Now, we use Proposition 4.2.2 to see that extremal Lancret helices in the Heisenberg group are those whose torsion satisfies

$$\tau^2 = r^2(4\cos^2\varphi - 3) = \frac{r^2(m^2-3)}{m^2+1},$$

and, according to (4.3), they are geodesics in Hopf cylinders over circles with radius

$$R = \frac{1}{\kappa_g} = \frac{m}{r\left(\sqrt{(m^2+1)(m^2-3)} + m^2 - 1\right)}, \qquad m = \cot\varphi.$$

Then, for any $\varphi \in (0, \pi/6)$ we obtain the extremal Lancret helices parameterized by

$$\begin{aligned}\gamma_m(s) &= X\left(\frac{s}{\sqrt{1+m^2}}, \frac{ms}{\sqrt{1+m^2}}\right)\\ &= \left(R\cos\frac{s}{R\sqrt{1+m^2}}, R\sin\frac{s}{R\sqrt{1+m^2}}, \frac{(rR+m)s}{\sqrt{1+m^2}}\right).\end{aligned}$$

Now, in order to provide examples of extremals in the second class, that is, extremals with horizontal binormal, we can use the following alternative algorithm obviously derived from the results in Sect. 4.3.2. It works on any $\mathrm{E}(c,r)$ and, in particular, on the Heisenberg group:

(i) Consider any unit speed curve $\beta(u)$ in $\mathrm{B}^2(c)$ with curvature function $\kappa_g(u)$.
(ii) Define functions $\psi(u) = -\mathrm{arcot}\frac{\kappa_g(u)}{2r}$ and $s(u) = \int_0^u \csc\psi(v)\,dv$.
(iii) Now, in the Hopf cylinder $S_\beta = \pi^{-1}(\beta(u(s)))$, choose the curve $\gamma(s)$ with slope $\psi(u(s))$ to obtain an extremal with horizontal binormal in $\mathrm{E}(c,r)$.

Following this procedure we give a few more examples of critical curves belonging to the second class.

(1) An Extremal Built on a Catenary By solving the natural equations one gets a parameterization of the unit speed curve, unique up to motions in $\mathbb{R}^2$, determined by the curvature function

$$\kappa_g(u) = \frac{1}{1+u^2}.$$

The trace of this curve is a catenary which, therefore, can be parameterized as

$$\beta(u) = (x(u), y(u)) = (\operatorname{arcsinh}(u), \sqrt{1+u^2}).$$

Using this and integrating (4.9) one has

$$z(u) = r\left(\sqrt{1+u^2}\operatorname{arcsinh}(u) - u\right),$$

which means that a horizontal lifting, $\bar{\beta}$, of β is given by

$$\bar{\beta}(u) = \left(\operatorname{arcsinh}(u), \sqrt{1+u^2}, r\left(\sqrt{1+u^2}\operatorname{arcsinh}(u) - u\right)\right).$$

Now, it is easy to write a parameterization $X(u,t)$ for the Hopf cylinder shaped on β, $S_\beta = \pi^{-1}(\beta)$

$$X(u,t) = \left(\operatorname{arcsinh}(u), \sqrt{1+u^2}, r\left(\sqrt{1+u^2}\operatorname{arcsinh}(u) - u\right) + t\right).$$

Finally, in the flat surface S_β we choose the curve $\gamma(s)$ whose slope, measured with respect to the vertical axis, is given by

$$\psi(u) = -\operatorname{arcot}\frac{1}{2r(1+u^2)}.$$

A parameterization of $\gamma(s)$ is

$$\gamma(s) = X(u(s), t(s)),$$

where

$$u(s) = -\int_0^s \frac{2r(1+v^2)}{\sqrt{1+4r^2(1+v^2)^2}}\,dv, \qquad t(s) = \int_0^s \frac{1}{\sqrt{1+4r^2(1+v^2)^2}}\,dv,$$

which can be expressed in terms of Elliptic functions. Then, $\gamma(s)$ provides an extremal for the total curvature action (with horizontal binormal) on the Heisenberg group (Fig. 4.1a).

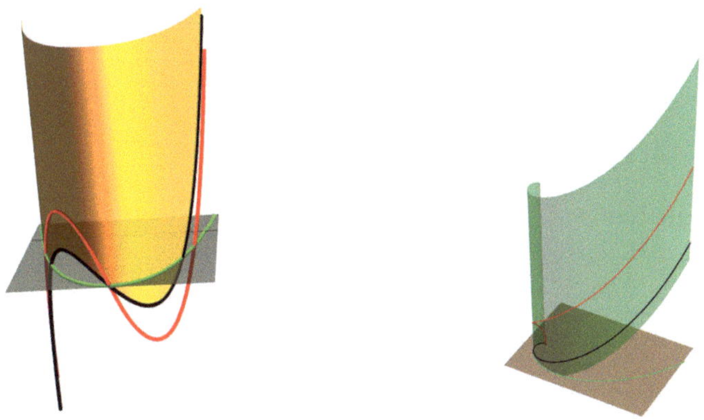

(a) β is a catenary. *(b) β is a logarithmic spiral.*

Fig. 4.1 A critical curve γ (curve in red) for the total curvature in Nil_3 with $r = 1$. Here, γ lies on a Hopf cylinder shaped on a planar curve β (curve in green), whose horizontal lifting $\bar{\beta}$ is shown in black. (**a**) β is a catenary. (**b**) β is a logarithmic spiral

(2) An Extremal Built on a Logarithmic Spiral Proceeding analogously, we see that a unit speed parameterization of the planar curve (unique up to motions in $\mathbb{R}^2$) determined by the curvature function

$$\kappa_g(u) = \frac{1}{u} \, ,$$

is given by $\beta(u) = (x(u), y(u))$, where

$$x(u) = \frac{u}{2} \left(\sin(\log(u)) + \cos(\log(u))\right) \, , \quad y(u) = \frac{u}{2} \left(\sin(\log(u)) - \cos(\log(u))\right) \, . \tag{4.10}$$

Moreover, solving (4.9) we obtain

$$z(u) = \frac{ru^2}{4} \, . \tag{4.11}$$

Hence, using (4.10) and (4.11) one sees that a horizontal lifting of β, $\bar{\beta}(u)$, and the Hopf cylinder shaped on β, $S_\beta = \pi^{-1}(\beta)$, can be explicitly and respectively parameterized by

$$\bar{\beta}(s) = (x(s), y(s), z(s)) \, , \quad X(u,t) = (x(u), y(u), z(u) + t) \, .$$

Again, in S_β we take the curve $\gamma(s)$ whose slope, measured with respect to the vertical axis, is given by

$$\psi(u) = -\mathrm{arcot}\frac{1}{u}\,.$$

A parameterization of $\gamma(s)$ is

$$\gamma(s) = X(u(s), t(s))\,,$$

where

$$u(s) = -\int_0^s \frac{2rv}{\sqrt{1+4r^2v^2}}\,dv = \frac{1-\sqrt{1+4r^2s^2}}{2r}\,,$$
$$t(s) = \int_0^s \frac{1}{\sqrt{1+4r^2v^2}}\,dv = \frac{1}{2r}\mathrm{arcsinh}(2rs).$$

Then, $\gamma(s)$ is also an extremal for the total curvature action (with horizontal binormal) on the Heisenberg group (Fig. 4.1b).

(3) An Extremal Constructed Out of an Involute of a Circle As a final example, let us consider the unit speed planar curve, β, determined by the curvature function

$$\kappa_g(u) = \frac{1}{\sqrt{u}}\,.$$

Solving the natural equation we obtain the following unit speed parameterization of the involute of a circle $\beta(u) = (x(u), y(u))$, where

$$x(u) = \sqrt{u}\sin(2\sqrt{u}) + \frac{1}{2}\cos(2\sqrt{u})\,, \quad y(u) = \frac{1}{2}\sin(2\sqrt{u}) - \sqrt{u}\cos(2\sqrt{u})\,. \tag{4.12}$$

Using these values of $x(u)$ and $y(u)$ in (4.9) and solving, we have

$$z(u) = r\sqrt{u}\,. \tag{4.13}$$

Formulae (4.12) and (4.13) allow us to obtain a parameterization of both, a horizontal lifting of β, $\bar{\beta}(u)$, and of the Hopf cylinder shaped on β, $S_\beta = \pi^{-1}(\beta)$, which can be explicitly given by

$$\bar{\beta}(s) = (x(s), y(s), z(s))\,, \quad X(u,t) = (x(u), y(u), z(u)+t)\,.$$

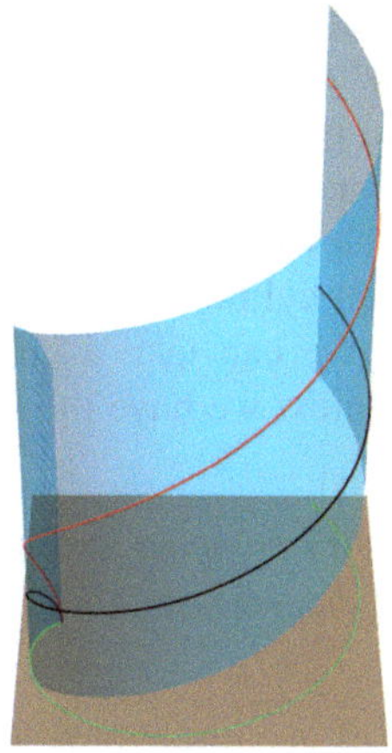

Fig. 4.2 A critical curve γ (curve in red) for the total curvature in Nil_3 with $r = 1$. Here, γ lies on a Hopf cylinder shaped on an involute of a circle β (curve in green), whose horizontal lifting $\bar{\beta}$ is shown in black

Then, the curve $\gamma(s)$ parameterized by $\gamma(s) = X(u(s), t(s))$, with

$$u(s) = \int_0^s \frac{2r\sqrt{v}}{\sqrt{1+4r^2 v}}\, dv = \frac{1}{4r^2}\left(2r\sqrt{s(1+4r^2 s)} - \ln(\sqrt{1+4r^2 s} + 2r\sqrt{s})\right),$$

$$t(s) = \int_0^s \frac{1}{\sqrt{1+4r^2 v}}\, dv = \frac{1}{2r^2}(\sqrt{1+4r^2 s} - 1),$$

lies in S_β forming an angle ψ with respect to the vertical axis given by

$$\psi(u) = -\operatorname{arcot}\frac{1}{2r\sqrt{u}}.$$

Therefore, $\gamma(s)$ is another extremal for the total curvature action (with horizontal binormal) on the Heisenberg group (Fig. 4.2).

4.5 Closed Extremals and Critical Levels

When one studies problems related with the differential geometry of curves, besides the solving natural equations problem, a second fundamental and open problem appears. It is called the closed curve problem. In this section we will focus on the solution to this problem for extremals of the total curvature in rotationally symmetric homogeneous 3-spaces. However, since, as we have shown, the space of extremals is made up of two families whose members are quite different from a qualitative point of view, we will treat the closed curve problem separately in both families (see [16]).

Let us consider first the family of extremals which are Lancret helices, that is, those with horizontal normal. According to Proposition 4.3.2, every $\mathrm{E}(c, r)$, except $\mathbb{H}^2(c) \times \mathbb{R}$, admits a real one-parameter class of Lancret helices providing extremals. In addition, we also know that these extremals are realized as geodesics of Hopf cylinders built over circles in $\mathtt{B}^2(c)$. Consequently, in order to get closed solutions, it is necessary that Hopf cylinders become tori, which happens just when the orbits close up. Otherwise, we can take a quotient of $\mathrm{E}(c, r)$ via a suitable vertical translation ϕ_t. In this context we have the following quantization principle à la Dirac.

Theorem 4.5.1 *Every* $\mathrm{E}(c, r)$ *with compact orbits, except* $\mathbb{H}^2(c) \times \mathbb{S}^1$*, admits a rational one-parameter class of closed Lancret helices which are extremals of the total curvature energy.*

Proof Since $\pi : \mathrm{E}(c, r) \to \mathtt{B}^2(c)$ is a circle bundle, then for every closed curve β in $\mathtt{B}^2(c)$ its Hopf cylinder, S_β, is actually a torus, which is embedded when the curve has no self-intersections. Moreover, it is a flat torus whose isometry type is known to be described as follows (see [2] and [13] for special cases).

First, note that if $r = 0$, then $\mathrm{E}(c, 0)$ is the Riemannian product $\mathbb{S}^2(c) \times \mathbb{S}^1$, where we have assumed, without loss of generality, that the orbits are unit circles. Since the holonomy is trivial, if β is a simple closed curve in $\mathbb{S}^2(c)$, with length L, then S_β is a flat torus generated by a rectangular lattice in the Euclidean plane, namely, it is $\mathbb{R}^2/\Gamma$, where Γ is the lattice generated by $\{(0, 2\pi), (L, 0)\}$. Then a Lancret helix (a geodesic in S_β) closes up if and only if its slope is a rational multiple of $2\pi/L$. Now the class of extremal Lancret helices in $\mathbb{S}^2(c) \times \mathbb{S}^1$ is the real one-parameter class of curves $\{\gamma_m : m \neq 0\}$, where γ_m is the geodesic with slope m in the Hopf torus built on the circle β_m in $\mathbb{S}^2(c)$, whose curvature is given by (4.7) with $r = 0$. As a consequence, the class of closed Lancret helices that are extremals of the total curvature in $\mathbb{S}^2(c) \times \mathbb{S}^1$ is the rational one-parameter family

$$\left\{\gamma_m : m = \frac{2\pi}{L} q, \quad q \in \mathbb{Q}\right\}.$$

If $r \neq 0$, then the holonomy is not trivial, which implies that the lattice describing the isometry type of S_β is not rectangular. In fact, it is generated by $\{(0, 2\pi), (L, 2A)\}$, where A is the area enclosed by β in $\mathtt{B}^2(c)$. Now a straight line $\alpha(s)$ with slope m provides a closed geodesic in the Hopf torus if and only if there exists $s_o > 0$ such that $\alpha(s_o) \in \Gamma$, which is equivalent to the following quantization formula

$$m = \frac{2}{L}(A + \pi\, r\, q), \qquad q \in \mathbb{Q}.$$

As above, let γ_m be the geodesic (Lancret helix) in the Hopf torus constructed on the circle, in $\mathtt{B}^2(c)$, with curvature given by (4.7). Then γ_m is an extremal of the total curvature energy if and only if

$$m^2 > 3 - \frac{c}{r^2}.$$

Consequently, the class of closed Lancret helices which are extremals coincides with the following rational one-parameter family

$$\left\{\gamma_m \,:\, m^2 > 3 - \frac{c}{r^2}, \quad m = \frac{2}{L}(A + \pi\, r\, q), \;\; q \in \mathbb{Q}\right\}.$$

□

Finally, we turn our attention to the case of extremals with horizontal binormal.

Theorem 4.5.2 *For every convex closed curve α in the Euclidean plane, there exists a closed curve γ in* $\mathrm{E}(c, r)$ *with the same curvature function as α and torsion $\tau = r$, which is an extremal, with horizontal binormal, of the total curvature action on* $\mathrm{E}(c, r)$.

Proof It follows from the proof of Theorem 4.3.10 for closed curves in the Euclidean plane. □

Remark 4.5.3

(1) The above theorem provides, up to congruences, a map from the space of convex closed curves in the Euclidean plane in the space of closed extremals with horizontal binormal in $\mathrm{E}(c, r)$. Even more, it is a one-to-one correspondence between both spaces of curves. In fact, if the Euclidean curve α is not closed then its image γ also is not, because S_β is not a torus. Observe that the profile curve has curvature function $\kappa_g(s) = -2r \cot \psi(s)$.

(2) If we consider the total curvature energy on closed curves in $\mathrm{E}(c, r)$, then, except in $\mathbb{H}^2(c) \times \mathbb{S}^1$, we obtain two families of extremals. On the one hand, a rational one-parameter class, $\mathbf{F}_N$, of Lancret helices (equivalently, curves with horizontal normal). On the other hand, the family of extremals, $\mathbf{F}_B$, with horizontal binormal, which has been identified with the class of convex closed curves in the Euclidean plane. The first family is discrete and consequently the critical values in $\mathbf{F}_N$ are quantized à la Dirac. The second family is certainly far from being discrete. However, it should be noted that according to the above one-to-one correspondence, a curve in $\mathbf{F}_B$ and its partner in the Euclidean plane have the same curvature function in the corresponding spaces. Consequently the critical values of the total curvature energy on closed extremals with horizontal binormal are not arbitrary, but they are integer multiples of a fixed value. Namely, the critical levels are

$$\mathcal{F}(\gamma) = \int_\gamma \kappa(s)\, ds = 2\pi\, \mathrm{i}(\alpha),$$

$\mathrm{i}(\alpha)$ standing for the rotation index of the corresponding closed curve in the Euclidean plane. This proves an interesting principle of quantization for the total curvature energy acting on closed curves in $\mathrm{E}(c, r)$.

Chapter 5
Lancret Helices in the Minkowski Space $\mathbb{L}^3$

We know that a Lancret helix in the Euclidean space $\mathbb{R}^3$ is a curve whose tangent indicatrix is a plane curve. The stright line perpendicular to that plane is called the axis of the helix.

We will now deal with Lancret helices in 3-dimensional Lorentzian space forms. A non-null curve γ embedded in $\mathbb{L}^3$ will be called a Lancret helix when its tangent indicatrix is contained in a plane $\Pi \subset \mathbb{L}^3$. It is known that Π can be degenerate or non-degenerate, depending on its induced metric is indefinite or Riemannian, respectively. Therefore, we will consider both cases separately, calling degenerate and non-degenerate Lancret helices, respectively. We will establish a Lancret theorem for Lancret helices in $\mathbb{L}^3$ which, formally, will coincide with the classic one in $\mathbb{R}^3$. Actually, we will prove that "the Lancret helices in $\mathbb{L}^3$ correspond to non-null curves in $\mathbb{L}^3$ for which the ratio from curvature to torsion is constant".

Regarding the behavior of the Lancret helices in both geometries, Euclidean and Lorentzian, we will highlight a substantial difference: while the Lancret helices in $\mathbb{R}^3$ are geodesics in right cylinders, the Lancret helices in $\mathbb{L}^3$ are geodesics either in right cylinders, or in B-flat scrolls, depending on the helices are non-degenerate or degenerate, respectively. This remarkable difference between both geometries confirms, once again, the important role of B-scrolls.

To extend the concept of the Lancret helix to the 3-dimensional de Sitter $\mathbb{S}^3_1$ and anti De Sitter $\mathbb{H}^3_1$ spaces, we will use the concept of a Killing field along a curve in a 3-dimensional space form, following the idea of Langer and Singer (see [69]). Lancret type results for $\mathbb{S}^3_1$ and $\mathbb{H}^3_1$ state deep differences between pseudospherical and pseudohyperbolic cases. The latter is analogous to that of $\mathbb{L}^3$, while in the former case we will see that the only Lancret helices will be the trivial ones. From this point of view, the role played by non-flat Lorentzian space forms, $\mathbb{S}^3_1$ and $\mathbb{H}^3_1$, corresponds to the same one for non-flat Riemannian cases $\mathbb{S}^3$ and $\mathbb{H}^3$, respectively (see [6]).

A. Ferrandez Izquierdo, *Helices*, RSME Springer Series 16,
https://doi.org/10.1007/978-3-032-09663-0_5

The study of Lancret helices in Lorentzian environments is supported by the interplay between geometry and integrable Hamiltonians systems (see [10, 67] and [68]).

5.1 Defining Lancret Helices in $\mathbb{L}^3$

We already know that Lancret helices in Euclidean space $\mathbb{R}^3$ are curves that form a constant angle with a fixed direction, the axis of the helix. Furthermore, it is clear that this idea does not make sense in the Lorentz-Minkowski space $\mathbb{L}^3$. Indeed, we cannot define the angle, including the hyperbolic angle [15], when a fixed lightlike (or null) direction is chosen as the axis. However, the constant slope for curves in $\mathbb{R}^3$ can be equivalently formulated as follows: *a curve has a constant slope and is therefore a Lancret helix if and only if its tangent indicatrix lies in a plane of* $\mathbb{R}^3$.

The tangent indicatrix can be defined for non-null curves in $\mathbb{L}^3$. Therefore, we can say that a non-null curve in $\mathbb{L}^3$ is a Lancret helix if its tangent indicatrix lies in a plane, [18]. It is clear that this property is equivalent to the fact that the acceleration vector field of the curve, when it is arc length parameterized, is orthogonal to a certain vector, ξ, in $\mathbb{L}^3$. So we can say that the line generated by such a vector, or simply ξ, is the axis of the Lancret helix. This allows us to distinguish between two types of non-null helices in $\mathbb{L}^3$, *degenerate* and *non-degenerate* according to whether the axis is null or not null, respectively.

Now, the above definition does not make sense for null curves since we cannot define the tangent indicatrix of such a curves. Therefore, we have to find another way to define a Lancret helix, naturally equivalent to that given, to avoid causality. To do so, we first observe that in $\mathbb{R}^3$, a curve is a Lancret helix if and only if it can be seen as a geodesic in a right cylinder built on a plane curve (cross section or directrix). According to a classical result of Hartman and Niremberg, [52], cylinders over plane curves form the whole class of complete and flat surfaces in $\mathbb{R}^3$.

The entire family of embedded flat Lorentzian surfaces (also called timelike) in $\mathbb{L}^3$ was obtained by L. K. Graves [48]. Up to motions in $\mathbb{L}^3$, this family is made up of two subfamilies:

- On one hand, the class $\mathcal{C}_1$ of Lorentzians right cylinders built on non-null curves. It is clear that this family does not present any surprise since it is made up of surfaces that keep a certain parallelism with the Euclidean cylinders. They are essentially the flat ruled surfaces whose generatrices are non-null straight lines.
- On the other hand, a class, $\mathcal{C}_2$, of surfaces that have no meaning in a Euclidean context. They are flat ruled surfaces with null generators (we will call them *null scrolls*), which, as we shall see, is equivalent to their mean curvature function vanishes identically.

We will then see that Lancret helix theory in $\mathbb{L}^3$ presents a canon similar to that of Euclidean ones, appearing as geodesics of flat surfaces. More precisely, we will prove the following results

- *Up to motions in $\mathbb{L}^3$, the degenerate Lancret helices are precisely the geodesics of the surfaces in class $\mathcal{C}_1$.*
- *Up to motions in $\mathbb{L}^3$, the non-degenerate Lancret helices are precisely the geodesics of the surfaces in class $\mathcal{C}_2$.*

This program provides an alternative way to define the notion of Lancret helix regardless the causal character of the curve. *A curve in $\mathbb{L}^3$, no matter its causal character, it will be said a Lancret helix if it can be seen as a geodesic on a flat surface.*

Later we will see that any null curve in $\mathbb{L}^3$ can be suitably parameterized to view it as a geodesic of a flat *null scroll*. Actually, we will see that there are infinitely many ways to do so, and therefore we can say that the null curves in $\mathbb{L}^3$ are Lancret helices with infinite axes.

5.2 Lorentzian Surfaces and Null Scrolls

A key difference between Euclidean and Lorentzian geometries is the existence of Lorentzian surfaces in $\mathbb{L}^3$. These surfaces, being non degenerate, behave quite differently from Euclidean surfaces. For example, any Lorentzian surface $\mathbf{S} \subset \mathbb{L}^3$ can be locally parameterized with two families of lightlike, or null, curves. Indeed, let $\mathbf{X}(u, v)$ be a local parameterization of $\mathbf{S}$ where the induced metric is given by

$$\mathbf{G} = \begin{pmatrix} g_{11} & g_{12} \\ g_{12} & g_{22} \end{pmatrix}.$$

Then the lightlike curves satisfy the condition

$$\begin{pmatrix} u' & v' \end{pmatrix} \begin{pmatrix} g_{11} & g_{12} \\ g_{12} & g_{22} \end{pmatrix} \begin{pmatrix} u' \\ v' \end{pmatrix} = 0.$$

As $\det(\mathbf{G}) < 0$, the above equation can be decomposed in linear factors as follows

$$(M\,u' + N\,v')(M\,u' + P\,v') = 0,$$

where

$$M^2 = g_{11}, \quad M(N + P) = 2\,g_{12}, \quad N\,P = g_{22},$$

so that

$$M\,u' + N\,v' = 0 \quad \text{and} \quad M\,u' + P\,v' = 0$$

give us two different vector fields having lightlike integral curves which are the coordinate curves of the asked parameterization.

5.3 Ruled Surfaces in $\mathbb{L}^3$

A ruled surface $\mathbf{S}(\gamma, B) \subset \mathbb{L}^3$ is defined by

$$\mathbf{X}(s,t) = \gamma(s) + t\,B(s), \qquad s \in I_1, \quad t \in I_2,$$

where $I_1, I_2 \subset \mathbb{R}$ are intervals and $\gamma = \gamma(s) \subset \mathbb{L}^3$ is a curve usually referred *base curve* or *directrix*, and $B(s)$ is a transversal vector field along γ, *generatrix flow*. We can distinguish two big families of ruled Lorentzian surfaces according to the generatrix flow be lightlike or not. The usual metric will be $\langle (a,b,c), (x,y,z) \rangle = ax + by - cz$.

Ruled Surfaces with Non-null Flow
Consider the following class of ruled surfaces

$$\mathcal{F}_1 = \{\mathbf{S}(\gamma, B) \,:\, \langle B(s), B(s) \rangle \neq 0\}.$$

For any surface in this class we can choose, if necessary, a non-null base curve [1]. In fact, reparameterizing the base curve and normalizing, in necessary, the generatrix flow, we can assume that

$$\langle B(s), B(s) \rangle = \varepsilon = \pm 1, \qquad \langle \gamma'(s), B(s) \rangle = -1,$$

as well as the surface is Lorentzian, otherwise it wouldn't make sense, so that

$$g(s,t) = \varepsilon \left(2t \langle \gamma'(s), B'(s) \rangle + t^2 \langle B'(s), B'(s) \rangle \right) - 1 < 0.$$

We are looking for a curve $\beta(s) = \gamma(s) + t(s)B(s)$ such that $\langle \beta'(s), \beta'(s) \rangle = \varepsilon$ and $\beta'(s)$ and $B(s)$ are linearly independent. All of this is equivalent to the function $t(s)$ being a solution of the following differential equation

$$(t'(s))^2 - 2\varepsilon t'(s) + g(s,t) = 0,$$

whose discriminant is positive and therefore it is integrable. Furthermore, the linear independence between $\beta'(s)$ and $B(s)$ follows from $\langle \beta'(s), \beta'(s) \rangle = \langle B(s), B(s) \rangle = \varepsilon$ and $\langle \beta'(s), B(s) \rangle = -1 + \varepsilon t'(s) \neq \pm\varepsilon$.

Surfaces in this class exhibit similar behavior to that of surfaces in a Euclidean context. Here are some examples of them:

Example 5.3.1

(1) **First kind Lorentzian helicoid.** Defined by

$$\mathbf{X}(s,t) = \gamma(s) + t\,B(s) = (0,s,0) + \frac{t}{\sqrt{1-c^2 s^2}}\,(cs,0,1).$$

(2) **Second kind Lorentzian helicoid.** Defined by

$$\mathbf{X}(s,t) = \gamma(s) + t\,B(s) = (0,s,0) + \frac{t}{\sqrt{c^2 s^2-1}}\,(1,0,cs).$$

(3) **Third kind Lorentzian helicoid.** Defined by

$$\mathbf{X}(s,t) = \gamma(s) + t\,B(s) = (s,0,0) + \frac{t}{\sqrt{1+c^2 s^2}}\,(0,-cs,1).$$

(4) **Enneper's conjugate surface of second kind.** Defined by

$$\mathbf{X}(s,t) = \gamma(s) + t\,B(s) = (0,s,0) + \frac{t}{\sqrt{1-2cs}}\,(cs,1,1-cs).$$

Ruled Surfaces with Null Flow or *Null Scrolls*

Consider now the class of ruled surfaces generated by a lightlike generatrix flow

$$\mathcal{F}_2 = \{\mathbf{S}(\gamma,B) \,:\, \langle B(s), B(s)\rangle = 0\}.$$

We use a similar argument as above to see that, if necessary, we can always take a lightlike base curve. Otherwise, we can find a unit reparameterization of the base curve and normalize the generatrix flow, so we get

$$\langle \gamma'(s), \gamma'(s)\rangle = \varepsilon = \pm 1, \qquad \langle \gamma'(s), B(s)\rangle = -1.$$

Now we look for a lightlike curve $\beta(s) = \gamma(s) + t(s)B(s)$ which is transversal to $B(s)$. To do that the function $t(s)$ should be a solution of the Ricatti differential equation

$$2t'(s) = \varepsilon + 2t(s)\langle \gamma'(s), B'(s)\rangle + t^2(s)\langle B'(s), B'(s)\rangle. \tag{5.1}$$

It is also clear that $\langle \beta'(s), B(s)\rangle = \langle \gamma'(s), B(s)\rangle \neq 0$. From now on the surfaces in this family will be called *null scrolls* [48]. Here we exhibit some examples of them.

Example 5.3.2

(1) Consider the *null scroll* $\mathbf{S}(\gamma, B)$ parameterized by $\mathbf{X}(s,t) = \gamma(s) + t\,B(s)$, where

$$\gamma(s) = \left(\frac{(2+\log s)^2 - c^2}{4s}, -\frac{c(2+\log s)}{2s}, -\frac{(2+\log s)^2 + c^2}{4s}\right),$$

is a lightlike curve and

$$B(s) = \frac{s}{2c^2}\left(c^2 - \log^2 s, 2\,c\log s, c^2 + \log^2 s\right), \quad c \neq 0,$$

is a lightlike vector field along γ.

(2) Let $\mathbf{S}(\gamma, B)$ be the *null scroll* parameterized by $\mathbf{X}(s,t) = \gamma(s) + t\,B(s)$, where

$$\gamma(s) = \left(\frac{e^s}{2}(\cos s + \sin s), \frac{e^s}{2}(-\cos s + \sin s), e^s\right)$$

is a lightlike curve and

$$B(s) = e^{-s}\,(-\sin s, \cos s, 1)$$

is the corresponding lightlike generatrix flow along γ.

5.4 Computing the Curvatures of Null Scrolls

We know that any *null scroll*, $\mathbf{S}(\gamma, B) : \mathbf{X}(s,t) = \gamma(s) + t\,B(s)$, can be parameterized in terms of a lightlike base curve

$$\langle \gamma'(s), \gamma'(s)\rangle = \langle B(s), B(s)\rangle = 0 \qquad \langle \gamma'(s), B(s)\rangle = -1.$$

The induced metric from $\mathbb{L}^3$ is given by

$$\begin{pmatrix} g_{11} & g_{12} \\ g_{21} & g_{22} \end{pmatrix} = \begin{pmatrix} 2t\,\langle \gamma'(s), B'(s)\rangle + t^2\,\langle B'(s), B'(s)\rangle & -1 \\ -1 & 0 \end{pmatrix}.$$

Hence, it is a Lorentzian surface. To compute its Gauss map and the corresponding shape operator, let us define $C(s) = \gamma'(s) \times B(s)$, which is a spacelike unit vector field along $\gamma(s)$ and orthogonal to $\mathbf{S}(\gamma, B)$. Then it defines the Gauss map along the base curve. The Gauss map is given by

$$N(s,t) = \frac{X_s(s,t) \times X_t(s,t)}{|g_{11}g_{22} - g_{12}^2|} = C(s) + t\,B'(s) \times B(s).$$

A simple computation yields

$$B'(s) \times B(s) = -f(s)\, B(s),$$

where $f(s) = \langle \gamma'(s), B'(s) \times B(s) \rangle = \det[\gamma'(s), B'(s), B(s)]$, and so we get

$$N(s,t) = -t\, f(s)\, B(s) + C(s).$$

Sometimes the function $-f(s)$ is referred to the *distribution parameter* of the ruled surface $\mathbf{S}(\gamma, B)$.

As for the computation of the shape operator, dN, of a *null scroll*, we can see that

$$N_s = -f\, X_s - \left(\langle \gamma', \gamma'' \times B \rangle + t\, f'\right) X_t$$

and

$$N_s = -f\, X_t,$$

so that the correspong matrix with respect to the chosen reference is

$$dN \equiv \begin{pmatrix} f & t\, f' + \langle \gamma', \gamma'' \times B \rangle \\ 0 & f \end{pmatrix}.$$

The, the mean, $\mathbf{H}$, and Gauss, $\mathbf{K}$, curvatures of the *null scroll* $\mathbf{S}(\gamma, B)$ are given by

$$\mathbf{H}(s,t) = f(s) \qquad \text{and} \qquad \mathbf{K}(s,t) = f(s)^2.$$

Here we list some properties of the peculiar geometry of these surfaces in $\mathbb{L}^3$:

1. The mean curvature, up to the sign, is nothing but the distribution parameter.
2. Since $\mathbf{H}^2(s,t) = \mathbf{K}(s,t)$, the *null scrolls* are Weingarten surfaces (see [35]). Actually, they are the only Lorentzian Weingarten surfaces in $\mathbb{L}^3$ whose curvature functions are related as above [35] and [58].
3. A *null scroll* is maximal ($\mathbf{H} = 0$) if and only if is flat ($\mathbf{K} = 0$).
4. Each *null scroll* is a Willmore surface in the conformal class of $\mathbb{L}^3$ [7].
5. The *null scrolls* provide solutions to the 2-dimensional non-linear sigma model and simetry $\mathbf{O}(2,1)$ [7].

Example 5.4.1 The *null scroll* built on the curve $\gamma(s) = (\cos s, \sin s, s)$, with lightlike flow $B(s) = (h(s), 0, h(s))$, being h a differentiable function, is given by

$$\mathbf{X}(s,t) = (\cos s + t\, h(s), \sin s, 1 + t\, h(s)).$$

To parameterize this *null scroll* on a null base curve, we need to obtain a solution of the Ricatti differential equation (5.1), which in this case becomes into a linear differential equation, because $B'(s)$ still is lightlike

$$t'(s) = \frac{\varepsilon}{2} - h'(s) \sin s \, t(s).$$

Now, the general solution of this differential equation is

$$t(s) = e^{\int h'(s) \sin s \, ds} \left[\frac{\varepsilon}{2} \int e^{-\int h'(s) \sin s \, ds} + C \right], \quad C = \text{constant}.$$

This provides a one parameter class of lightlike curves that works as base curves for the given *null scroll.*

5.5 Lancret Helices in the 3-Dimensional Minkowski Space

Following the classical terminology of the Euclidean geometry (see, for instance, [75]) we will say that γ is a *Lancret, or generalized, helix* in $\mathbb{L}^3$ if its tangent indicatrix lies in a plane of $\mathbb{L}^3$. That means that there exists a vector $v \neq 0$ in $\mathbb{L}^3$ which is orthogonal to the acceleration vector field of γ. The straight line generated by v is uniquely determined and we will refer to it as the *axis* of γ. In particular, we will say that a Lancret helix is *degenerate* or *non-degenerate* according to its axis is null or non-null, respectively.

It is obvious that non-null curves in $\mathbb{L}^3$ with zero torsion are examples of non-degenerate Lancret helices. In fact, such a curve lies in a non-degenerate 2-plane in $\mathbb{L}^3$ and a unit vector in $\mathbb{L}^3$ orthogonal to this plane works as the axis of the Lancret helix.

Now given a Lancret helix γ in $\mathbb{L}^3$ with axis v, we can define a translation vector field $\widetilde{V}$ in $\mathbb{L}^3$ by $\widetilde{V}_p = v$, for any $p \in \mathbb{L}^3$. Let V be $\widetilde{V}$ restricted to γ. Then V defines a Killing vector field along γ with constant length, i.e., $\langle V, V \rangle$ is constant, and orthogonal to the acceleration vector field of γ.

Assume now that W is a Killing vector field along a non-null curve γ, with constant length and orthogonal to its normal vector field N. From (1.4a) we can write $W = aT + bB$, a and b being constants. Now use (1.4b) to get $\overline{\nabla}_T W = \lambda N$, where $\lambda = \varepsilon_2(a\kappa + b\tau)$ is constant. Finally, Eq. (1.4c) yields $\lambda\tau(\tau/k)' = 0$. From here we consider the following cases.

(i) $\tau \equiv 0$. Then γ is a a non-degenerate Lancret helix. It is not difficult to see that $W = B$ (unless γ is a circle which will be considered next) and so it extends to a translation vector field $\widetilde{W}$ in $\mathbb{L}^3$.

(ii) *k and τ both are constant.* Then γ is a helix. Now the Killing vector field W is not uniquely determined. In fact, for any couple of contants a and b, in the rectifying plane we define the vector field $W(s) = aT + bB$, which works as a Killing vector field along γ. In spite of that, we can determine a Killing vector field along γ, say $V(s)$, being parallel along γ and thus it extends to a translation vector field $\widetilde{V}(s)$ such that $\widetilde{V}(s) = v \in \mathbb{L}^3$. Indeed just choose a and b such that $a\kappa + b\tau = 0$. Therefore γ is a non-degenerate Lancret helix, unless $\varepsilon_2 = 1$ (which means that N is spacelike or the rectifying plane is Lorentzian anywhere) and $\tau = \pm\kappa$, and then γ is degenerate.

(iii) $\lambda = 0$. Then W is a uniquely determined Killing vector field along γ. Furthermore, it is parallel along γ and extends to a translation vector field $\widetilde{W}$ such that $\widetilde{W} = v \in \mathbb{L}^3$. Therefore, γ is a Lancret helix whose axis is v (or W). And γ is degenerate when W is null, which yields $\varepsilon_2 = 1$ and $\tau = \pm\kappa$.

We will refer to curves in the first two classes as trivial Lancret helices.
Summarizing we have the following (see [11]).

Theorem 5.5.1 (The Lancret Theorem in $\mathbb{L}^3$) *Let γ be a non-null immersed curve in $\mathbb{L}^3$ with curvature and torsion functions κ and τ, respectively. Then the following statements are equivalent:*

(a) *γ is a Lancret helix in $\mathbb{L}^3$;*
(b) *There exists a constant length Killing vector field V along γ which is orthogonal to the acceleration vector field of γ;*
(c) *There exists a constant r such that $\tau = r\kappa$.*

Moreover a Lancret helix γ is degenerate if and only if $r = \pm 1$ and its normal vector field is spacelike. The Killing vector field V in (b) is not uniquely determined if γ is a helix (κ and τ both are constant); however, in this case, V can be uniquely determined, up to constants, once it is chosen parallel along γ (in other words, its extended Killing vector field in $\mathbb{L}^3$ is a translation vector field).

Theorem 5.5.2 (Solving Natural Equation for Non-degenerate Lancret Helices) *Let β be a non-null immersed curve in $\mathbb{L}^3$. Then β is a non-degenerate Lancret helix if and only if it is a geodesic in some right cylinder whose directrix and generatrix are both non-null.*

Proof Let v be a unit vector in $\mathbb{L}^3$ and α a unit speed curve in a plane orthogonal to v. The Frenet equations of α are

$$\begin{aligned} \overline{\nabla}_{\overline{T}}\overline{T} &= \delta_2\bar{\kappa}\overline{N}, \\ \overline{\nabla}_{\overline{T}}\overline{N} &= -\delta_1\bar{\kappa}\overline{T}, \end{aligned} \tag{5.2}$$

where $\{\overline{T}, \overline{N}\}$ is the Frenet frame along α, $\bar{\kappa}$ its curvature function and δ_1, δ_2 the causal characters of $\overline{T}$ and $\overline{N}$, respectively. Notice that the causal character of v is $-\delta_1\delta_2$.

Let us consider the right cylinder $C_{\alpha,v}$ in $\mathbb{L}^3$ generated by α and v, which is naturally parameterized as $X(s,t) = \alpha(s) + tv$. It is well known that the geodesics of $C_{\alpha,v}$ are the images under X of straight lines in the (s,t)-plane. Choose such a geodesic $\gamma(s) = \alpha(s) + msv$, where m is a certain constant. Then the translation field $\widetilde{V}$ in $\mathbb{L}^3$ determined by v induces a Killing vector field along γ with constant length and orthogonal to the acceleration vector $\gamma''(s) = \alpha''(s)$. Since v is non-null, Theorem 5.5.1 implies that γ is a non-degenerate Lancret helix.

Conversely, suppose β is a non-degenerate Lancret helix. Then there exists a certain constant r such that $\tau = r\kappa$ (of course κ and τ denote the curvature and torsion functions of β). One can also choose a unit vector, say v, lying on the axis of β. We take a (non-degenerate) plane P in $\mathbb{L}^3$ which is orthogonal to v. Up to congruences in P, there exists a unique curve in P, say α, with curvature function $\bar{\kappa} = |\beta'|\kappa$ and $\delta_2 = \varepsilon_2$ (notice that the causal character δ_1 of α is determined by δ_2 and the causal character of v). Let $C_{\alpha,v}$ be the right cylinder generated by α and v. Then it is parameterized by $X(s,t) = \alpha(s) + tv$. Finally we choose the geodesic of $C_{\alpha,v}$ defined by $\gamma(s) = \alpha(s) + msv$, where $m = \delta_1\varepsilon_3 r$. Then γ is a non-null geodesic, because $\delta_2 = -1$ provided that $m^2 = 1$. Finally, it is easy to see that γ and β have the same curvature and torsion functions, as well as the same causal characters. This concludes the proof. □

Theorem 5.5.3 (Solving Natural Equation for Degenerate Lancret Helices) *Let β be a non-null immersed curve in $\mathbb{L}^3$. Then β is a degenerate Lancret helix if and only if it is a geodesic in some flat B-scroll in $\mathbb{L}^3$.*

Proof Let $\alpha(s)$ be a null curve in $\mathbb{L}^3$ with Cartan frame $\{A, B, C\}$ and $S_{\alpha,B}$ the flat B-scroll (i.e., $w_0 = 0$) parameterized by $X(s,t) = \alpha(s)+tB$. We choose a non-null geodesic of $S_{\alpha,B}$, say $\gamma(u) = \alpha(s(u))+t(u)B(s(u))$. Then the translation field $\tilde{B}$ in $\mathbb{L}^3$ determined by B induces a Killing vector field along γ, also denoted by B, with constant length and such that $\langle\gamma'(u), B\rangle = -s'(u)$ is constant, because the geodesic γ is the image under X of a straight line. Therefore, from Theorem 5.5.1, γ is a degenerate Lancret helix in $\mathbb{L}^3$.

To prove the converse, let β be a degenerate Lancret helix that we parameterize with constant speed, say $\langle\beta', \beta'\rangle = p$ constant. From the theorem of Lancret we know that the curvature τ and torsion κ functions of β agree (we can change orientation if necessary) and the acceleration vector field of β is spacelike, i.e., $\varepsilon_2 = 1$. We define the following vector fields

$$A = \frac{|\beta'|}{2}(T + B),$$

$$B = -\frac{\varepsilon_1}{|\beta'|}(T - B),$$

$$C = N,$$

where $\{T, N, B\}$ is the Frenet frame along β, $|\beta'| = \sqrt{\varepsilon_1 p}$ and ε_1 denotes, as usual, the causal character of β.

Let α be a curve in $\mathbb{L}^3$ with tangent vector field A, then α is a null curve in $\mathbb{L}^3$. Furthermore, $\{A, B, C\}$ is a Cartan frame along α with $w_0 = 0$ and $\rho = \kappa|\beta'|$ (see Sect. 5.2). Let $S_{\alpha,B}$ be the corresponding flat B-scroll, which is parameterized by $X(s,t) = \alpha(s) + tB$. Finally, choose the geodesic in $S_{\alpha,B}$ given by $\gamma(s) = \alpha(s) + msB$, where $m = -p/2$. It is not difficult to see that γ and β have the same curvature and torsion functions, and also the same causal character, showing that they are congruent in $\mathbb{L}^3$. $\square$

Chapter 6
Lancret Helices in the de Sitter $\mathbb{S}_1^3$ and Anti de Sitter $\mathbb{H}_1^3$ Spaces

6.1 Setup

Let $\mathbb{R}_t^{n+2}$ be the $(n+2)$-dimensional pseudo-Euclidean space with index t endowed with the indefinite inner product given by

$$\langle x, y\rangle = -\sum_{i=1}^{t} x_i y_i + \sum_{j=t+1}^{n+2} x_j y_j,$$

where $(x_1, \ldots, x_{n+2})$ is the usual coordinate system. Let $\mathbb{S}_\nu^{n+1} = \{x \in \mathbb{R}_\nu^{n+2} : \langle x, x\rangle = 1\}$ and $\mathbb{H}_{\nu+1}^{n+1} = \{x \in \mathbb{R}_{\nu+1}^{n+2} : \langle x, x\rangle = -1\}$ be the unit pseudo-sphere and the unit pseudo-hyperbolic space, respectively. They are pseudo-Riemannian hypersurfaces of index ν in $\mathbb{R}_\nu^{n+2}$ and $\mathbb{R}_{\nu+1}^{n+2}$, respectively, with constant sectional curvature $c = +1$ and $c = -1$, respectively. Let M denote $\mathbb{S}_1^3$, $\mathbb{H}_1^3$ or $\mathbb{L}^3$ according to $c = +1, c = -1$ or $c = 0$, respectively, and E will stand for the pseudo-Euclidean space where M is lying.

Let p be a point in M and $v \in T_pM$ a tangent vector. Then v is said to be *spacelike*, *timelike* or *null* according to $\langle v, v\rangle > 0$, $\langle v, v\rangle < 0$, or $\langle v, v\rangle = 0$ and $v \neq 0$, respectively. Notice that the vector $v = 0$ is spacelike. The category into which a given tangent vector falls is called its causal character. These definitions can be generalized for curves as follows. A curve γ in M is said to be *spacelike* if all of its velocity vectors γ' are spacelike; similarly for timelike and null.

For a better understanding of the next construction we will bring back the notion of cross product in the tangent space T_pM at any point p in M. There is a natural orientation in T_pM defined as follows: an ordered basis $\{X, Y, Z\}$ in T_pM is positively oriented if $\det[pXYZ] > 0$, where $[pXYZ]$ is the matrix with p, X, Y, Z as row vectors. Now let ω be the volumen element on M defined by $\omega(X, Y, Z) = \det[pXYZ]$. Then given $X, Y \in T_pM$, the cross product $X \times Y$ is the unique vector in T_pM such that $\langle X \times Y, Z\rangle = \omega(X, Y, Z)$, for any $Z \in T_pM$.

A. Ferrandez Izquierdo, *Helices*, RSME Springer Series 16,
https://doi.org/10.1007/978-3-032-09663-0_6

A *non-null curve* $\gamma(s)$ in M is said to be a unit speed curve if $\langle\gamma'(s), \gamma'(s)\rangle = \varepsilon$ (ε being $+1$ or -1 according to γ is spacelike or timelike, respectively). A unit speed curve $\gamma(s)$ in M, s being the arc length parameter, is called a *Frenet curve* if it admits a Frenet frame field $\{T = \gamma', N, B\}$, where $B = T \times N$, satisfying the Frenet equations

$$\begin{aligned}
\overline{\nabla}_T T &= \varepsilon_2 \kappa N, \\
\overline{\nabla}_T N &= -\varepsilon_1 \kappa T - \varepsilon_3 \tau B, \\
\overline{\nabla}_T B &= \varepsilon_2 \tau N,
\end{aligned}$$

where ε_1, ε_2 and ε_3 denote the causal characters of T, N and B, respectively (in particular, $\varepsilon_i = \pm 1$ and $\varepsilon_1\varepsilon_2\varepsilon_3 = -1$), $\overline{\nabla}$ is the semi-Riemannian connection on M and $\kappa = \kappa(s)$ and $\tau = \tau(s)$ are the curvature and the torsion functions of γ, respectively.

The unit tangent vector field $T = \gamma'$ defines a mapping from γ to $Q = \{q \in E : \langle q, q\rangle = \pm 1\}$, which is usually called the *tangent indicatrix* of γ and, from now on, it will be also denoted by T.

Now let $\alpha(s)$ be a null curve in M with Cartan frame $\{A, B, C\}$, i.e., A, B, C are vector fields tangent to M along $\alpha(s)$ satisfying the following conditions:

$$\langle A, A\rangle = \langle B, B\rangle = 0, \quad \langle A, B\rangle = -1,$$

$$\langle A, C\rangle = \langle B, C\rangle = 0, \quad \langle C, C\rangle = 1,$$

and

$$\begin{aligned}
\dot{\alpha} &= A, \\
\dot{A} &= \rho C, \qquad \rho = \rho(s) \neq 0, \\
\dot{B} &= c\alpha + w_0 C, \qquad w_0 \text{ being a constant}, \\
\dot{C} &= w_0 A + \rho B.
\end{aligned}$$

If we consider the immersion $X : (s, t) \to \alpha(s) + tB(s)$, then X defines a Lorentz surface, with constant Gaussian curvature $c + w_0^2$, that L. K. Graves [48] called a *B-scroll*. An easy computation shows that the unit normal vector is given, up to the sign, by $\xi(s, t) = w_0 t B(s) + C(s)$.

6.2 Lancret Helices in Non-flat 3-Dimensional Lorentzian Space Forms

In order to generalize the notion of Lancret helix to 3-dimensional Lorentzian spaces of nonzero constant curvature, we profit by Theorem 5.5.1. A curve γ in M is said to be a Lancret helix if there exists a Killing vector field V along γ with constant length and orthogonal to the acceleration vector field of γ. We will say that V is an axis of the Lancret helix γ. Obvious examples of Lancret helices in M are the following. Curves with torsion vanishing anywhere, where the unit binormal works as an axis. Helices are also Lancret helices, where any vector field chosen in the rectifying plane having constant coordinates relative to T and B runs as an axis.

We can follow notation and terminology used in $\mathbb{L}^3$ to say that zero torsion curves are non-degenerate Lancret helices, because the axis B is obviously non-null. As for curves with both constant curvature and torsion we know that for any pair of constants a and b the vector field along γ given by $V(s) = aT + bB$ is always a Killing vector field. Of course, when $\varepsilon_2 = -1$, i.e., the rectifying plane is positive definite at any point, all Killing vector fields $V(s)$ are non-null and we will say that the Lancret helix is non-degenerate. However, if $\varepsilon_2 = 1$, i.e., the rectifying plane is Lorentzian, we have Killing vector fields along γ being either spacelike, or timelike, or null. It does not allow us to decide if such a Lancret helix is degenerate or not. However, we can determine a unique Killing vector field along the helix by forcing it to be parallel along γ. The helix is said to be *degenerate* or *non-degenerate* according to V is null or non-null, respectively.

Let $\gamma(s)$ be a Lancret helix in M with curvature $\kappa > 0$. Let $V(s)$ be an axis and assume, without loss of generality, that $\langle V, V\rangle = \varepsilon$, where $\varepsilon = -1, 0, 1$. From Eq. (1.4a) we deduce that

$$V(s) = fT(s) + hB(s), \quad \text{and} \quad \varepsilon = \varepsilon_1 f^2 + \varepsilon_3 h^2, \tag{6.1}$$

for certain constants f and h. By using the Frenet equations of γ we get

$$\overline{\nabla}_T V = \varepsilon_2 (f\kappa + h\tau) N, \tag{6.2}$$

and

$$\overline{\nabla}^2_T V = -\varepsilon_1\varepsilon_2\kappa(f\kappa + h\tau)T + \varepsilon_2(f\kappa' + h\tau')N - \varepsilon_2\varepsilon_3\tau(f\kappa + h\tau)B. \tag{6.3}$$

Now from Eqs. (1.4b), (6.1) and (6.3) we deduce that $f\kappa' + h\tau' = 0$ from which we get

$$\tau = b\kappa + a, \tag{6.4}$$

for certain constants a and b. On the other hand, from (6.3), jointly with the Frenet equations of γ, we obtain

$$\overline{\nabla}_T^3 V = -\varepsilon_1\varepsilon_2\lambda\kappa' T - \lambda(\varepsilon_1\kappa^2 + \varepsilon_3\tau^2)N - \varepsilon_2\varepsilon_3\lambda\tau' B, \tag{6.5}$$

where λ stands for the constant $f\kappa + h\tau$. Now Eq. (1.4c), jointly with Eqs. (6.1)–(6.5), yields

$$\tau(\lambda\tau'\kappa - \lambda\kappa'\tau - ch\kappa') = 0,$$

and then

$$h\kappa'\tau(a^2 + c) = 0. \tag{6.6}$$

In particular, the above equation shows that in the De Sitter space $\mathbb{S}_1^3$, $(c = +1)$, the only Lancret helices are the two classes described just before this discussion. So we have prove the following result (see [11]).

Theorem 6.2.1 (The Lancret Theorem in the De Sitter Space) *A non-null immersed curve γ in $\mathbb{S}_1^3$ is a Lancret helix if and only if either*

(1) *$\tau \equiv 0$ and γ is a curve in some totally geodesic surface of $\mathbb{S}_1^3$; or*
(2) *γ is a helix in $\mathbb{S}_1^3$ (i.e. curvature κ and torsion τ both are constant).*

Furthermore, Lancret helices of the first type have only one axis (the binormal) which is parallel and so they are non-degenerate. In contrast, Lancret helices of the second type have a plane (the rectifying plane) of axes. However they only have a parallel axis. This axis is null, and so the Lancret helix is degenerate if and only if $\varepsilon_2 = +1$ (the normal vector is spacelike) and $\tau = \pm\kappa$; otherwise, the helix is non-degenerate.

In the anti De Sitter space, besides the two classes of trivial Lancret helices, we have another class. This kind of Lancret helices can be characterized from Eqs. (6.5) and (6.6), where $c = -1$, as the curves in $\mathbb{H}_1^3$ whose curvature and torsion are related by

$$\tau = b\kappa \pm 1,$$

for a certain constant b. These Lancret helices admit only one axis $V = fT + hB$, which is defined by

$$\frac{f}{h} = -b = \frac{1-\tau}{\kappa}.$$

The causal character of this axis is

$$\varepsilon = h^2\left(\varepsilon_1\frac{(\tau-1)^2}{\kappa^2} + \varepsilon_3\right).$$

In particular, a Lancret helix of this type is degenerate if and only if $\varepsilon_2 = 1$ and $b = \pm 1$.

Summarizing we have shown the following.

Theorem 6.2.2 (The Lancret Theorem in the Anti De Sitter Space) *A non-null immersed curve γ in $\mathbb{H}_1^3$ is a Lancret helix if and only if either*

(1) $\tau \equiv 0$ *and γ is a curve in some totally geodesic surface of $\mathbb{H}_1^3$. The curve admits only one axis which agrees with its binormal, being parallel along the curve and non-null. The Lancret helix is non-degenerate; or*
(2) *γ is a helix in $\mathbb{H}_1^3$. It admits a plane (the rectifying plane) of axes but only one is parallel along γ. This parallel axis is null, and so γ is degenerate, if and only if $\varepsilon_2 = +1$ and $\tau = \pm\kappa$. Otherwise, γ is non-degenerate; or*
(3) *there exists a certain constant b such that the curvature κ and the torsion τ functions of γ are related by $\tau = b\kappa \pm 1$. The curve admits a unique axis which can not be parallel along γ. It is null, and so γ is degenerate, if and only if $b = \pm 1$ and γ has spacelike normal vector ($\varepsilon_2 = +1$).*

Remark 6.2.3 Compare Theorems 6.2.1 and 6.2.2 with Theorems 1 and 3 in [6], respectively.

6.3 Solving the Natural Equations

In [10] we have just constructed a new class of submanifolds in $\mathbb{H}_1^3(-1)$ defined by means of two semi-Riemannian submersions $\pi_s : \mathbb{H}_1^3(-1) \to \mathbb{H}_s^2(-4)$, $s = 0, 1$ (see details therein). By pulling back via π_s a non-null curve γ in $\mathbb{H}_s^2(-4)$ we get the total horizontal lifting of γ, which is an immersed flat surface M_γ in $\mathbb{H}_1^3(-1)$, that will be called the *semi-Riemannian Hopf cylinder associated to γ*. Notice that if $s = 0$, M_γ is a Lorentzian surface, whereas if $s = 1$, M_γ is Riemannian or Lorentzian, according to γ be spacelike or timelike, respectively.

Let $\gamma : I \to \mathbb{H}_s^2(-4)$ be a unit speed curve with Frenet frame $\{\bar{T}, \bar{N}\}$ and curvature function $\bar{\kappa}$. Let $\bar{\gamma}$ be a horizontal lifting of γ to $\mathbb{H}_1^3(-1)$ with Frenet frame $\{T, N, B\}$, curvature $\kappa = \bar{\kappa} \circ \pi_s$ and torsion $\tau = 1$. Recall that B is nothing but the unit tangent vector field to the fibers along $\bar{\gamma}$. Then the Hopf cylinder M_γ can be orthogonally parameterized by

$$X(t, z) = \begin{cases} \cos(z)\bar{\gamma}(t) + \sin(z)B(t) \text{ when } s = 0, \\ \cosh(z)\bar{\gamma}(t) + \sinh(z)B(t) \text{ when } s = 1. \end{cases}$$

Notice that a unit normal vector field to M_γ into $\mathbb{H}_1^3(-1)$ is obtained from the complete horizontal lifting of $\bar{N}$ and it is, of course, N along each horizontal lifting of γ. As a consequence we have that M_γ is a flat surface with mean curvature function α is given by $\alpha = \frac{1}{2}\kappa$ (see [11]).

Theorem 6.3.1 (Solving Natural Equation for Non-degenerate Lancret Helices in $\mathbb{H}_1^3(-1)$) *Let β a non-null immersed curve in $\mathbb{H}_1^3(-1)$. Then β is a non-degenerate Lancret helix if and only if it is a geodesic in some Hopf cylinder M_γ.*

Proof Let $\beta(s)$ be an arclength parameterized geodesic in M_γ, then there exists two constants a and b such that

$$T(s) = \beta'(s) = aX_t + bX_z,$$

with $\varepsilon_1 a^2 + \varepsilon_3 b^2 = \delta_1$, δ_1 being the causal character of β. A direct computation shows that the curvature ρ and the torsion τ of β satisfy

$$\rho = \varepsilon_2 a^2 \kappa + 2ab,$$
$$\tau^2 = \varepsilon_2 \rho^2 - \varepsilon_1 \delta_1 \kappa \rho + 1.$$

It is not difficult to see that $\tau = r\rho \pm 1$, $r = b/a$, showing that β is a Lancret helix. Moreover, if the normal vector N is spacelike, then $r \neq 1$ and then β is non-degenerate.

To prove the converse, let β be a non-degenerate Lancret helix in $\mathbb{H}_1^3(-1)$ with curvature ρ and torsion τ. Then there exists a constant r (with $r \neq \pm 1$ if the normal vector to β is spacelike) such that $\tau = r\rho \pm 1$. We choose $\varepsilon_1 = \pm 1$ and s in $\{0, 1\}$ in order to $\delta_1(\varepsilon_1 - (-1)^s r^2)$ be positive, δ_1 being the causal character of β. Let γ be the unique curve, up to motions, in $\mathbb{H}_2^1(-4)$ with curvature $\bar{\kappa} = \delta_1((-1)^s - \varepsilon_1 r^2)\rho - 2\varepsilon_1(-1)^s r$ and causal character defined by ε_1. Let α be the geodesic in the Hopf cylinder M_γ given by $\alpha(s) = X(as, bs)$ with

$$a^2 = \frac{\delta_1}{\varepsilon_1 - (-1)^s r^2} \qquad \text{and} \qquad b^2 = r^2 a^2.$$

It is easy to see that β and α have the same curvature and torsion, and also the same causal character, showing that they are congruent. □

Theorem 6.3.2 (Solving Natural Equation for Degenerate Lancret Helices in $\mathbb{H}_1^3(-1)$) *Let β a non-null immersed curve in $\mathbb{H}_1^3(-1)$. Then β is a degenerate Lancret helix if and only if it is a geodesic in some flat B-scroll over a null curve.*

Proof Let $\beta(u)$ be a geodesic of some flat B-scroll $S_{\alpha,B}$ in $\mathbb{H}_1^3(-1)$ (i.e., $w_0 = \pm 1$) parameterized by $\beta(u) = \alpha(s(u)) + t(u)B(s(u))$. Then the normal vector to β in $\mathbb{H}_1^3(-1)$ is given by $N(u) = \beta(u) - \alpha(s(u)) + C(s(u))$. From here we obtain that $\overline{\nabla}_T N = T + s'(u)\rho B$. By using the Frenet equations for β we deduce that the vector $(1 + \varepsilon_1 \kappa)T + \varepsilon_3 \tau B$ is null, where κ y τ stand for the curvature and torsion of β, respectively. Therefore N is spacelike and $\tau = \pm\varepsilon_1 \kappa \pm 1$, which proves that β is a degenerate Lancret helix.

Conversely, let β be a curve in $\mathbb{H}^3_1(-1)$ with curvature κ and torsion τ satisfying that $\tau = \kappa + \varepsilon_1$ and the normal vector of β is spacelike (the other cases are similar). We define the null curve α in $\mathbb{H}^3_1(-1)$ by the equation

$$\alpha(s) = \beta(s) - \frac{1}{2}s(T(s) - B(s)),$$

and the following vector fields

$$\begin{aligned} A(s) &= -\frac{\varepsilon_1}{2}s\beta(s) + \frac{1}{2}(T(s) + B(s)) + \frac{\varepsilon_1}{2}sN(s), \\ B(s) &= -\varepsilon_1(T(s) - B(s)), \\ C(s) &= -\frac{1}{2}s(T(s) - B(s)) + N(s). \end{aligned}$$

It is not difficult to see that $\{A, B, C\}$ is a Cartan frame along α with $w_0 = 1$ and $\rho = \tau$. Let $S_{\alpha,B}$ be the B-scroll in $\mathbb{H}^3_1(-1)$ parameterized by $X(s,t) = \alpha(s) + tB(s)$. Then it is clear that $\beta(s) = X(s, -\frac{\varepsilon_1}{2}s)$ and so β is a geodesic of that B-scroll. □

Remark 6.3.3 It is worth noting that in the "if" part in Theorems 6.3.1 and 6.3.2 we have only used the existence of an axis, not necessarily parallel. Then if γ is a helix in $\mathbb{H}^3_1(-1)$ with Lorentzian rectifying plane anywhere, it turns out have both null and non-null axes. Therefore γ is a geodesic in a Hopf cylinder as well as in a flat B-scroll over a null curve.

Chapter 7
Lightlike Helices in $\mathbb{L}^n$, $\mathbb{S}_1^n$ and $\mathbb{H}_1^n$

The general theory of curves in a Riemannian manifold M have been developed long time ago and today we have a deep knowledge of both its local and global geometry. When M is a proper semi-Riemannian manifold (that is, the index ν of the metric of M satisfies $1 \leq \nu \leq \dim(M) - 1$) there exist three families of curves (spacelike, timelike, lightlike or null curves) depending on their causal characters. It is well-known, [82], that the study of timelike curves has many analogies and similarities with that of spacelike curves. However, the fact that the induced metric on a lightlike curve is degenerate leads to a much more complicated study and also different from the non-degenerate case, even sometimes the geometry of a lightlike hypersurface in a semi-Euclidean space can be investigated by using the geometry of the hypersurface as a Riemannian hypersurface in a Euclidean space (see [20]).

In the geometry of lightlike, or null, curves difficulties arise because the arc length vanishes, so that it is not possible to normalize the tangent vector in the usual way. A method of proceeding is to introduce a new parameter called the pseudo-arc (already used by Vessiot [104]) which normalizes the derivative of the tangent vector. This was the point of view followed by W. B. Bonnor in [24] where he defined two curvatures K_2 and K_3 in terms of the pseudo-arc and a third curvature K_1 which takes only two values, 0 and 1, according as the lightlike curve is a straight line or, otherwise, without any inflexion points (see also the work by M. Castagnino [30]). J. L. Synge, [95], followed a different procedure by supposing that in addition to the equation of the lightlike curve $\gamma(w)$, a lightlike vector $\mathbf{p}$ parallel to $\gamma'(w)$ is also given. Then there exists a unique parameter u such that $\mathbf{p} = \gamma'(u)$, and this new parameter allows us to study the geometry of the curve. From a physical point of view, this corresponds to specify both the world-line and the momentum of a photon.

The importance of the study of lightlike curves and its presence in the physic theories is clear from the fact (see [59]) that the classical relativistic string is a surface or world-sheet in Minkowski space which satisfies the Lorentzian analogue of the minimal surface equations. The string equations simplify to the wave

A. Ferrandez Izquierdo, *Helices*, RSME Springer Series 16,
https://doi.org/10.1007/978-3-032-09663-0_7

equation, plus a couple of extra simple equations, and by solving the 2-dimensional wave equation it turns out that strings are equivalent to pairs of lightlike curves, or a single lightlike curve in the case of an open string (see also [60–62, 93] and [102]). In [78], the authors show that there exists a geometrical particle model based enterely on the geometry of the lightlike curves in 4-dimensional Minkowskian spacetime which under quantization yields the wave equations corresponding to massive spinning particles of arbitrary spin. In [79], the authors consider the simplest geometrical particle model associated to lightlike curves in 3-dimensional Lorentz-Minkowski space and show that under quantization the action, which is proportional to the pseudo-arc length, yields the $(2+1)$-dimensional anyonic field equation supplemented with a Majorana-like relation on mass and spin, i.e. mass $\times$ spin $= \alpha^2$, with α the coupling constant in front of the action.

Motivated by the growing importance of lightlike, or null, curves in mathematical physics, A. Bejancu began in [19] an ambitious program to study the differential geometry of lightlike curves in Lorentzian manifolds and, more generally, in semi-Riemannian manifolds. From a complementary vector subbundle to the tangent bundle of a lightlike curve, he obtains the Frenet equations (with respect to a general Frenet frame) and proves certain theorems of existence and uniqueness for lightlike curves in Lorentzian manifolds (see also [37]).

The importance of helices and Lancret, or generalized, helices in physical theories is also well known, as one can see, for example, in [11] and references therein. This chapter is taken from [43].

7.1 Preliminaries

Let $(\overline{M}, g)$ be a proper $(m+2)$-dimensional semi-Riemannian manifold of index q and let us consider C a smooth curve in $\overline{M}$ locally parameterized by $\gamma : I \subset \mathbb{R} \to \overline{M}$. The curve C is said to be *lightlike or null* if the tangent vector $\gamma'(t)$ to C at any point is a lightlike vector.

The following concepts are taken from [37]. Let TC denote the tangent bundle of C and define, as in the non-generate case, the bundle $TC^{\perp}$ by

$$TC^{\perp} = \bigcup_{p\in C} T_pC^{\perp}; \quad T_pC^{\perp} = \{\xi_p \in T_pM : g(\xi_p, V_p) = 0\},$$

where V_p is a lightlike vector tangent to C at p. It is well known that $TC^{\perp}$ is of rank $m+1$. Since V_p is a lightlike vector, it easily follows that TC is a vector subbundle of $TC^{\perp}$ of rank 1. Then we may consider a complementary vector subbundle $S(TC^{\perp})$ to TC in $TC^{\perp}$ such that

$$TC^{\perp} = TC \perp S(TC^{\perp}),$$

where $\perp$ means orthogonal direct sum. It is known that the subbundle $S(TC^{\perp})$, called the *screen vector bundle* of C, is non-degenerate and of dimension m. Note that, in contrast with the non-degenerate case, the tangent bundle is contained in the normal bundle, and the screen bundle is not unique. These two properties leads to a much more difficult and also different geometry of lightlike curves with respect to non-degenerate (spacelike or timelike) curves.

Since $S(TC^{\perp})$ is non-degenerate, we have the decomposition $T\overline{M}|_C = S(TC^{\perp})\perp S(TC^{\perp})^{\perp}$, where $S(TC^{\perp})^{\perp}$ is the complementary orthogonal vector bundle to $S(TC^{\perp})$ in $T\overline{M}|_C$. The following result is well-known.

Theorem 7.1.1 ([19]) *Let C be a lightlike curve of a semi-Riemannian manifold $(\overline{M}, g)$ and consider $S(TC^{\perp})$ a screen vector bundle of C. Then there exists a unique vector bundle E over C, of rank 1, such that on each coordinate neighbourhood $U \subset C$ there is a unique section $N \in \Gamma(E|_C)$ satisfying*

$$\langle \gamma'(t), N\rangle = 1$$

and

$$\langle N, N\rangle = \langle N, X\rangle = 0, \quad \textit{for any } X \in \Gamma(S(TC^{\perp})|_U).$$

The above vector bundle E will be denoted by $ntr(C)$ and it is called the *lightlike transversal bundle* of C with respect to $S(TC^{\perp})$. The vector field N is called the *lightlike transversal vector field* of C with respect to $\gamma'(t)$. We define the transversal vector bundle of C, $tr(C)$, as the vector bundle

$$tr(C) = ntr(C)\perp S(TC^{\perp}),$$

and then we have

$$T\overline{M}|_C = TC \oplus tr(C) = (TC \oplus ntr(C))\perp S(TC^{\perp}), \tag{7.1}$$

from which the following result easily follows.

Proposition 7.1.2 ([37]) *Let C be a lightlike curve of a semi-Riemannian manifold $(\overline{M}, g)$ of index q. Then any screen vector bundle of C is semi-Riemannian of index $q - 1$. Hence, if M is a Lorentzian manifold, then any screen vector bundle is Riemannian.*

Let M_1^n be an orientable Lorentzian manifold and consider C a lightlike curve locally parameterized by $\gamma : I \subset \mathbb{R} \to M_1^n$. Assume that $\{\gamma', \gamma'', \dots, \gamma^{(n)}\}$ is a linearly independent family and define $E_i = \text{span}\{\gamma', \gamma'', \dots, \gamma^{(i)}\}$, $i = 1, \dots, n$. We denote by $E_i^{\perp}$ the complementary vector bundle to E_i in E_{i+1}, that is, $E_{i+1} = E_i \perp E_i^{\perp}$. Now we are going to construct a Frenet frame on $\overline{M}$ along C.

To do that, let $L \in E_1$, so that $\gamma' = \overline{k}_1 L$, for a certain function $\overline{k}_1$. Since $E_2 = E_1 \oplus \mathrm{span}\{\gamma''\}$, we have

$$\dim E_1^{\perp} = \dim E_2 - \dim E_1 + \dim(\mathrm{rad}(E_2) \cap E_1) = 1 + d_{21},$$

where we write $d_{ij} = \dim(\mathrm{rad}(E_i) \cap E_j)$. It is easy to see that $d_{21} = 1$ and then $E_1^{\perp} = E_2$, so that we can choose a unit spacelike vector W_1 satisfying $E_2 = \mathrm{span}\{L, W_1\}$. Now, since $E_3 = E_2 \oplus \mathrm{span}\{\gamma^{(3)}\}$ we obtain that $d_{32} = 0$ and then $E_2^{\perp} = E_1$. Since E_3 is a Lorentzian subspace of E_n, then there exists only one lightlike vector N such that $\langle L, N\rangle = 1$, $\langle W_1, N\rangle = 0$ and $E_3 = \mathrm{span}\{L, N, W_1\}$. In general, for $i = 2, \ldots, n-3$, we can find orthonormal spacelike vectors $\{W_1, \ldots, W_i\}$ such that $E_{i+2} = \mathrm{span}\{L, N, W_1, \ldots, W_i\}$ and the basis $\{\gamma', \gamma'', \ldots, \gamma^{(i+2)}\}$ and $\{L, N, W_1, \ldots, W_i\}$ have the same orientation. Finally, the vector W_m, $m = n - 2$, is chosen in order that the basis $\{L, N, W_1, \ldots, W_m\}$ is positively oriented. The vector bundle $\mathrm{span}\{W_1, \ldots, W_m\}$ is a screen vector bundle of C. An easy computation shows that there exist functions $\{\overline{k}_1, \ldots, \overline{k}_{m+3}\}$ such that the following equations hold (compare with [19])

$$\begin{aligned}
\gamma' &= \overline{k}_1 L,\\
L' &= \overline{k}_2 L + \overline{k}_3 W_1,\\
N' &= -\overline{k}_2 N + \overline{k}_4 W_1 + \overline{k}_5 W_2,\\
W_1' &= -\overline{k}_4 L - \overline{k}_3 N,\\
W_2' &= -\overline{k}_5 L + \overline{k}_6 W_3,\\
&\ \vdots\\
W_i' &= -\overline{k}_{i+3} W_{i-1} + \overline{k}_{i+4} W_{i+1},\\
&\ \vdots\\
W_{m-1}' &= -\overline{k}_{m+2} W_{m-2} + \overline{k}_{m+3} W_m,\\
W_m' &= -\overline{k}_{m+3} W_{m-1}.
\end{aligned} \tag{7.2}$$

The set $F = \{L, N, W_1, \ldots, W_m\}$ satisfying the above equations is called the *Frenet frame* on $\overline{M}$ along C with respect to the screen vector bundle $\mathrm{span}\{W_i\}$. The functions $\{\overline{k}_1, \ldots, \overline{k}_{m+3}\}$ are called the *curvature functions* of C with respect to F. Those equations are called the *Frenet equations* of C with respect to F.

7.2 The Cartan Frame of a Lightlike Curve

The goal of this section is to find a Frenet frame with the minimal number of curvatures and such that they are invariant under Lorentzian transformations. Without loss of generality we may assume that γ is parameterized by the pseudo-arc parameter, that is, $\langle \gamma'', \gamma'' \rangle = 1$. Now choose $L = \gamma'$ and $W_1 = \gamma''$, so that $\overline{k}_1 = 1, \overline{k}_2 = 0$ and $\overline{k}_3 = 1$.

It is not difficult to show that the lightlike transversal bundle is generated by the section

$$N = -\gamma^{(3)} - \frac{1}{2}\left\langle \gamma^{(3)}, \gamma^{(3)} \right\rangle \gamma',$$

and then the forth curvature is given by

$$\overline{k}_4 = \langle N', W_1 \rangle = \frac{1}{2}\left\langle \gamma^{(3)}, \gamma^{(3)} \right\rangle. \tag{7.3}$$

From the equation

$$\overline{k}_5 W_2 = N' - \overline{k}_4 W_1 = -\gamma^{(4)} - \left\langle \gamma^{(4)}, \gamma^{(3)} \right\rangle \gamma' - \left\langle \gamma^{(3)}, \gamma^{(3)} \right\rangle \gamma'',$$

we easily deduce

$$\begin{aligned} \overline{k}_5 &= \pm\sqrt{\langle \gamma^{(4)}, \gamma^{(4)} \rangle - \langle \gamma^{(3)}, \gamma^{(3)} \rangle^2}, \\ W_2 &= \mp\frac{1}{\overline{k}_5}\left(\gamma^{(4)} + \left\langle \gamma^{(3)}, \gamma^{(3)} \right\rangle \gamma'' + \left\langle \gamma^{(4)}, \gamma^{(3)} \right\rangle \gamma' \right). \end{aligned}$$

A direct computation shows that $W_1' = -\overline{k}_4 L - \overline{k}_3 N$. After rename the curvature functions ($k_1 = \overline{k}_4, k_2 = \overline{k}_5$ and $k_3 = \overline{k}_6$), the first four Frenet equations write down as follows:

$$\begin{aligned} L' &= W_1, \\ N' &= k_1 W_1 + k_2 W_2, \\ W_1' &= -k_1 L - N, \\ W_2' &= -k_2 L + k_3 W_3. \end{aligned}$$

If the dimension is greater than 5, then by a similar reasoning we have the following Frenet equations

$$\begin{aligned} W_i' &= -k_i W_{i-1} + k_{i+1} W_{i+1}, \qquad i \in \{3, \ldots, m-1\}, \\ W_m' &= -k_m W_{m-1}, \quad m = n-2. \end{aligned}$$

Then we have shown the following theorem.

Theorem 7.2.1 *Let $\gamma : I \to M^n_1$, $n = m + 2$, be a lightlike curve parameterized by the pseudo-arc such that $\{\gamma'(t), \gamma''(t), \dots, \gamma^{(n)}(t)\}$ is a basis of $T_{\gamma(t)}M^n_1$ for all t. Then there exists only one Frenet frame satisfying the equations*

$$\begin{aligned} L' &= W_1, \\ N' &= k_1 W_1 + k_2 W_2, \\ W_1' &= -k_1 L - N, \\ W_2' &= -k_2 L + k_3 W_3, \\ W_i' &= -k_i W_{i-1} + k_{i+1} W_{i+1} \quad i \in \{3, \dots, m-1\}, \\ W_m' &= -k_m W_{m-1}, \end{aligned} \tag{7.4}$$

and fullfilling

(i) *For $2 \leqslant i \leqslant m-1$, $\{\gamma', \gamma'', \dots, \gamma^{(i+2)}\}$ and $\{L, N, W_1, \dots, W_i\}$ have the same orientation.*
(ii) *$\{L, N, W_1, \dots, W_m\}$ is positively oriented.*

Definition 7.2.2 A lightlike curve in M^n_1 satisfying the conditions of the above theorem is called a *Cartan curve*. The above Frenet frame and curvatures $\{k_1, k_2, \dots, k_m\}$ are called the *Cartan frame* and the *Cartan curvatures*, respectively, of the curve γ.

Let $\mathcal{A}$ and $\mathcal{B}$ denote the references $\{\gamma', \gamma'', \dots, \gamma^{(n)}\}$ and $\{L, N, W_1, W_2, \dots, W_m\}$, respectively, and write A and B for the matrices of the metric with respect to $\mathcal{A}$ and $\mathcal{B}$, respectively. Let D_i denote the i-th order main determinant of the matrix A, that is,

$$D_i = \begin{vmatrix} \langle \gamma', \gamma' \rangle & \langle \gamma', \gamma'' \rangle & \cdots & \langle \gamma', \gamma^{(i)} \rangle \\ \langle \gamma'', \gamma' \rangle & \langle \gamma'', \gamma'' \rangle & \cdots & \langle \gamma'', \gamma^{(i)} \rangle \\ \vdots & \vdots & \ddots & \vdots \\ \langle \gamma^{(i)}, \gamma' \rangle & \langle \gamma^{(i)}, \gamma'' \rangle & \cdots & \langle \gamma^{(i)}, \gamma^{(i)} \rangle \end{vmatrix}$$

Proposition 7.2.3 *Let $\gamma : I \to M^n_1$ be a Cartan curve. Then the Cartan curvatures $\{k_1, k_2, k_3, \dots, k_m\}$ are given by*

$$k_1 = \frac{1}{2}\left\langle \gamma^{(3)}, \gamma^{(3)} \right\rangle, \quad k_2^2 = -D_4, \quad k_i^2 = \frac{D_i D_{i+2}}{D_{i+1}^2}.$$

Moreover, $k_2 < 0$, $k_i > 0$ for all $i \in \{3, \dots, m-1\}$, and $k_m > 0$ or $k_m < 0$ according to $\mathcal{A}$ is positively or negatively oriented, respectively.

Proof The formula for k_1 (old $\overline{k}_4$) was computed in (7.3). After a straightforward computation, it is not difficult to see that

$$\begin{pmatrix} \gamma' \\ \gamma'' \\ \gamma^{(3)} \\ \gamma^{(4)} \\ \gamma^{(5)} \\ \gamma^{(6)} \\ \vdots \\ \gamma^{(n)} \end{pmatrix} = \begin{pmatrix} 1 & 0 & 0 & 0 & 0 & 0 & \dots & 0 \\ 0 & 0 & 1 & 0 & 0 & 0 & \dots & 0 \\ -k_1 & -1 & 0 & 0 & 0 & 0 & \dots & 0 \\ -k_1' & 0 & -2k_1 & -k_2 & 0 & 0 & \dots & 0 \\ * & * & * & * & -k_2k_3 & 0 & \dots & 0 \\ * & * & * & * & * & -k_2k_3k_4 & \dots & 0 \\ \vdots & \vdots & \vdots & \vdots & \vdots & \vdots & \ddots & \vdots \\ * & * & * & * & * & * & \dots & -k_2k_3\dots k_m \end{pmatrix} \begin{pmatrix} L \\ N \\ W_1 \\ W_2 \\ W_3 \\ W_4 \\ \vdots \\ W_m \end{pmatrix}.$$

If we denote by P the above matrix, then from $A = P^t B P$, P^t standing for the transpose matrix, we get

$$\det A = (\det P)^2 \det B = -k_m^2 k_{m-1}^4 \dots k_2^{2m-2}.$$

On the other hand, a direct computation leads to

$$\begin{aligned} D_3 &= -1, \\ D_4 &= -k_2^2, \\ D_5 &= -k_3^2 k_2^4, \\ D_6 &= -k_4^2 k_3^4 k_2^6, \\ &\vdots \\ D_i &= -k_{i-2}^2 k_{i-3}^4 \dots k_2^{2(i-3)}, \\ &\vdots \\ D_{m+2} &= -k_m^2 k_{m-1}^4 \dots k_2^{2(m-1)}. \end{aligned}$$

Now the result is clear by using Theorem 7.2.1. □

Corollary 7.2.4 *The Cartan curvatures of a curve C in M_1^n are invariant under Lorentzian transformations.*

Definition 7.2.5 A curve is said to be a *helix* if it has constant Cartan curvatures.

A long and messy computation shows that if γ is a helix then it satisfies the following differential equation

$$\gamma^{(n+1)} = a_1\gamma' + a_2\gamma^{(3)} + \cdots + a_s\gamma^{(n-1)}, \quad \text{if } n \text{ is even, } n = 2s,$$

and

$$\gamma^{(n+1)} = a_1\gamma'' + a_2\gamma^{(4)} + \cdots + a_s\gamma^{(n-1)}, \quad \text{if } n \text{ is odd, } n = 2s + 1,$$

where the coefficients are given by

$$\begin{aligned} a_i = k_2^2 & \sum_{\substack{4\le j_1<\cdots<j_{s-i-1}\le n-2 \\ j_r-j_{r-1}\ge 2 \quad \forall r}} k_{j_1}^2 \cdots k_{j_{s-i-1}}^2 - 2k_1 \sum_{\substack{3\le j_1<\cdots<j_{s-i}\le n-2 \\ j_r-j_{r-1}\ge 2 \quad \forall r}} k_{j_1}^2 \cdots k_{j_{s-i}}^2 \\ & - \sum_{\substack{3\le j_1<\cdots<j_{s-i+1}\le n-2 \\ j_r-j_{r-1}\ge 2 \quad \forall r}} k_{j_1}^2 \cdots k_{j_{s-i+1}}^2, \qquad i = 1, \ldots, s-2, \end{aligned}$$

$$a_{s-1} = k_2^2 - 2k_1 \sum_{j=3}^{n-2} k_j^2 - \sum_{\substack{3\le j_1<j_2\le n-2 \\ j_2-j_1\ge 2}} k_{j_1}^2 k_{j_2}^2,$$

$$a_s = -2k_1 - \sum_{j=3}^{n-2} k_j^2.$$

In [24], Bonnor study and classify helices in $\mathbb{R}_1^4$. Now we are going to extend his results to $\mathbb{R}_1^5$, $\mathbb{S}_1^4$ and $\mathbb{H}_1^4$.

7.3 Lightlike Curves in $\mathbb{R}_1^n$

Definition 7.3.1 A basis $\mathcal{B} = \{L_1, N_1, \ldots, L_r, N_r, W_1, \ldots, W_m\}$ of $\mathbb{R}_q^n$, with $2r \le 2q \le n$ and $m = n-2r$, is said to be *pseudo-orthonormal* if it satisfies the following equations:

$$\langle L_i, L_j\rangle = \langle N_i, N_j\rangle = 0, \quad \langle L_i, N_j\rangle = \delta_{ij},$$
$$\langle L_i, W_\alpha\rangle = \langle N_i, W_\alpha\rangle = 0, \quad \langle W_\alpha, W_\beta\rangle = \varepsilon_\alpha\delta_{\alpha\beta},$$

where $i, j \in \{1, \ldots, r\}$, $\alpha, \beta \in \{1, \ldots, m\}$, $\varepsilon_\alpha = -1$ if $1 \le \alpha \le q-r$ and $\varepsilon_\alpha = 1$ if $q - r + 1 \le \alpha \le m$.

Lemma 7.3.2 *Let $\mathcal{B} = \{L_1, N_1, \ldots, L_r, N_r, W_1, \ldots, W_m\}$ be a basis of $\mathbb{R}_q^n$, with $2r \leq 2q \leq n$ and $m = n - 2r$. Consider $\mathcal{B}' = \{V_1, \ldots, V_q, V_{q+1}, \ldots, V_n\}$ where*

$$V_i = \begin{cases} \frac{1}{\sqrt{2}}(L_i - N_i) & i = 1, \ldots, r \\ W_{i-r} & i = r+1, \ldots, q \\ \frac{1}{\sqrt{2}}\left(L_{i-q} + N_{i-q}\right) & i = q+1, \ldots, q+r \\ W_{i-2r} & i = q+r+1, \ldots, n \end{cases} \tag{7.5}$$

The following conditions are equivalent:

(i) $\mathcal{B}$ is a pseudo-orthonormal basis.
(ii) $\mathcal{B}'$ is an orthonormal basis.
(iii) $\mathcal{B}'$ satisfies

$$-\sum_{\alpha=1}^{q} V_{\alpha i} V_{\alpha j} + \sum_{\beta=q+1}^{n} V_{\beta i} V_{\beta j} = \eta_{ij}.$$

(iv) $\mathcal{B}$ satisfies

$$\sum_{\alpha=1}^{r}\left(L_{\alpha i} N_{\alpha j} + L_{\alpha j} N_{\alpha i}\right) - \sum_{\beta=1}^{q-r} W_{\beta i} W_{\beta j} + \sum_{\theta=q-r+1}^{m} W_{\theta i} W_{\theta j} = \eta_{ij}.$$

Here $V_{\rho k}, L_{\rho k}, N_{\rho k}$ and $W_{\rho k}$ stand for the components of vectors V_ρ, L_ρ, N_ρ and W_ρ, respectively, and (η_{ij}) denotes the matrix of the canonical metric in the standard coordinates.

Proof (i) $\Leftrightarrow$ (ii) It is obvious.
(ii) $\Leftrightarrow$ (iii) Consider the matrices $V = \left(V_{ij}\right)$, $B = \left(b_{ij}\right)$ and $C = \left(c_{ij}\right)$ in $\mathcal{M}_{n\times n}(\mathbb{R})$ given by

$$b_{ij} = \langle V_i, V_j\rangle,$$
$$c_{ij} = -\sum_{\alpha=1}^{q} V_{\alpha i} V_{\alpha j} + \sum_{\beta=q+1}^{n} V_{\beta i} V_{\beta j}.$$

Put

$$V = \begin{pmatrix} A_1 & A_2 \\ A_3 & A_4 \end{pmatrix}, \quad B = \begin{pmatrix} B_1 & B_2 \\ B_3 & B_4 \end{pmatrix} \quad \text{and} \quad C = \begin{pmatrix} C_1 & C_2 \\ C_3 & C_4 \end{pmatrix},$$

where A_1, B_1 and C_1 are matrices in $\mathcal{M}_{q\times q}(\mathbb{R})$. Consider the complex matrix

$$A = \begin{pmatrix} A_1 & iA_2 \\ -iA_3 & A_4 \end{pmatrix} \in \mathcal{M}_{n\times n}(\mathbb{C}).$$

Then a straightforward computation shows that

$$AA^T = \begin{pmatrix} -B_1 & iB_2 \\ iB_3 & B_4 \end{pmatrix} \quad \text{and} \quad A^T A = \begin{pmatrix} -C_1 & -iC_2 \\ -iC_3 & C_4 \end{pmatrix}.$$

Then $\mathcal{B}'$ is orthonormal if and only if $C_1 = -I$, $C_4 = I$ and $C_2 = C_3 = 0$, and that concludes the proof.
(iii) $\Leftrightarrow$ (iv) From (7.5) we have

$$L_\alpha = \frac{1}{\sqrt{2}}\left(V_{\alpha+q} + V_\alpha\right) \quad \text{and} \quad N_\alpha = \frac{1}{\sqrt{2}}\left(V_{\alpha+q} - V_\alpha\right), \quad \alpha \in \{1, \ldots, r\},$$

and then

$$L_{\alpha i}N_{\alpha j} + N_{\alpha i}L_{\alpha j} = -V_{\alpha i}V_{\alpha j} + V_{(\alpha+q)i}V_{(\alpha+q)j}, \quad \alpha \in \{1, \ldots, r\},$$
$$i, j \in \{1, \ldots, n\},$$

which finishes the proof. □

Theorem 7.3.3 *Let $k_1, k_2, \ldots, k_m : [-\varepsilon, \varepsilon] \to \mathbb{R}$ be differentiable functions with $k_2 < 0$ and $k_i > 0$ for $i \in \{3, \ldots, m-1\}$. Let p be a point in $\mathbb{R}_1^n$, $n = m+2$, and consider $\{L^0, N^0, W_1^0, \ldots, W_m^0\}$ a positively oriented pseudo-orthonormal basis of $\mathbb{R}_1^n$. Then there exists a unique Cartan curve γ in $\mathbb{R}_1^n$, with $\gamma(0) = p$, whose Cartan frame $\{L, N, W_1, \ldots, W_m\}$ satisfies*

$$L(0) = L^0, N(0) = N^0, W_i(0) = W_i^0, \quad i \in \{1, \ldots, m\}.$$

Proof According to the general theory of differential equations, there exists a unique solution $\{L, N, W_1, \ldots, W_m\}$ of (7.4), defined on an interval $[-\varepsilon, \varepsilon]$, and satisfying the initial conditions of the theorem. A straightforward computation, bearing in mind (7.4), leads to

$$\frac{d}{dt}\left(L_i N_j + L_j N_i + \sum_{\alpha=1}^{m} W_{\mathbb{R}_i^\alpha} W_{\alpha j}\right) = 0, \quad i, j \in \{1, \ldots, n\}.$$

Since $\{L(0), N(0), W_1(0), \ldots, W_m(0)\}$ is a pseudo-orthonormal basis, then the above equation jointly with Lemma 7.3.2 implies

$$L_i(t)N_j(t) + L_j(t)N_i(t) + \sum_{\alpha=1}^{m} W_{\alpha i}(t)W_{\alpha j}(t) = \eta_{ij}, \quad \forall t \in [-\varepsilon, \varepsilon].$$

Then by using again Lemma 7.3.2 we deduce that $\{L, N, W_1, \ldots, W_m\}$ is a pseudo-orthonormal basis for all t, and this concludes the proof. □

The following result shows that the Cartan curvatures determine curves satisfying the nondegeneracy conditions stated in Theorem 7.2.1.

Theorem 7.3.4 *If two Cartan curves C and $\bar{C}$ in $\mathbb{R}^n_1$ have Cartan curvatures $\{k_1, \ldots, k_n\}$, where $k_i : [-\varepsilon, \varepsilon] \to \mathbb{R}$ are differentiable functions, then there exists a Lorentzian transformation of $\mathbb{R}^n_1$ which maps C into $\bar{C}$.*

7.3.1 Lightlike Helices in $\mathbb{R}^5_1$

The goal of this section is to classify the family of lightlike helices in $\mathbb{R}^5_1$. Before to do that, we present some examples. From the general equation, we know that a lightlike helix in $\mathbb{R}^5_1$ satisfies the following differential equation

$$\gamma^{(6)} + (2k_1 + k_3^2)\gamma^{(4)} - (k_2^2 - 2k_1k_3^2)\gamma'' = 0,$$

which will help us to find the examples.

Example 7.3.5 (Helices of Type 1) Let ω, σ and h be three non-zero constants such that $\frac{1}{\sigma^2} < h^2 < \frac{1}{\omega^2}$ and let $\gamma_{\omega,\sigma,h} : \mathbb{R} \to \mathbb{R}^5_1$ be the curve defined by

$$\gamma_{\omega,\sigma,h}(t) = \left(ht, \frac{1}{\omega}\sqrt{\frac{h^2\sigma^2 - 1}{\sigma^2 - \omega^2}} \sin \omega t, \frac{1}{\omega}\sqrt{\frac{h^2\sigma^2 - 1}{\sigma^2 - \omega^2}} \cos \omega t, \frac{1}{\sigma}\sqrt{\frac{1 - h^2\omega^2}{\sigma^2 - \omega^2}} \sin \sigma t, \right.$$
$$\left. \frac{1}{\sigma}\sqrt{\frac{1 - h^2\omega^2}{\sigma^2 - \omega^2}} \cos \sigma t \right).$$

Then it is easy to see that $\gamma_{\omega,\sigma,h}$ is a helix with curvatures given by

$$k_1 = \frac{1}{2}\left(\sigma^2 + \omega^2(1 - \sigma^2h^2)\right),$$
$$k_2^2 = -\omega^2\sigma^2\left(\omega^2h^2 - 1\right)\left(\sigma^2h^2 - 1\right),$$
$$k_3^2 = \omega^2\sigma^2h^2.$$

Example 7.3.6 (Helices of Type 2) Let ω, σ and h be three non-zero constants such that $0 < h^2\omega^2 < 1$ and let $\gamma_{\omega,\sigma,h} : \mathbb{R} \to \mathbb{R}_1^5$ be the curve defined by

$$\gamma_{\omega,\sigma,h}(t) = \left(\frac{1}{\omega}\sqrt{\frac{1+h^2\sigma^2}{\omega^2+\sigma^2}} \sinh \omega t, \frac{1}{\omega}\sqrt{\frac{1+h^2\sigma^2}{\omega^2+\sigma^2}} \cosh \omega t, \frac{1}{\sigma}\sqrt{\frac{1-h^2\omega^2}{\omega^2+\sigma^2}} \sin \sigma t, \right.$$
$$\left. \frac{1}{\sigma}\sqrt{\frac{1-h^2\omega^2}{\omega^2+\sigma^2}} \cos \sigma t, ht \right).$$

Then $\gamma_{\omega,\sigma,h}$ is a helix with curvatures given by

$$k_1 = \frac{1}{2}\left(\sigma^2 - \omega^2(1+\sigma^2h^2)\right),$$
$$k_2^2 = -\omega^2\sigma^2\left(\omega^2h^2 - 1\right)\left(\sigma^2h^2+1\right),$$
$$k_3^2 = \omega^2\sigma^2h^2.$$

Example 7.3.7 (Helices of Type 3) Let σ and h be two non-zero constants such that $0 < 2h^2 < 1$ and let $\gamma_{\sigma,h} : \mathbb{R} \to \mathbb{R}_1^5$ be the curve defined by

$$\gamma_{\sigma,h}(t) = \left(\frac{3}{2}\left(\frac{1-2h^2}{\sigma^2h^2}\right) t + \frac{1}{6}h^2t + t^3, \frac{h}{\sqrt{2}}t^2, \frac{3}{2}\left(\frac{1-2h^2}{\sigma^2h^2}\right) t - \frac{1}{6}h^2t + t^3, \right.$$
$$\left. \frac{\sqrt{1-2h^2}}{\sigma^2} \sin \sigma t, \frac{\sqrt{1-2h^2}}{\sigma^2} \cos \sigma t \right).$$

Then $\gamma_{\sigma,h}$ is a helix with curvatures given by

$$k_1 = \frac{1}{2}\sigma^2\left(1-2h^2\right),$$
$$k_2^2 = 2\sigma^4h^2\left(1-2h^2\right),$$
$$k_3^2 = 2\sigma^2h^2.$$

Theorem 7.3.8 *Let γ be a lightlike curve fully immersed in $\mathbb{R}_1^5$. Then γ is a helix if and only if it is congruent to a helix either of type 1, or type 2 or type 3.*

Proof Let k_1, k_2 and k_3 be the constant curvatures of γ such that $k_2 \neq 0 \neq k_3$. By the Congruence Theorem 7.3.4 it suffices to find a helix (of one of the above types) with these curvatures.

Case 1: Assume that $(2k_1 + k_3^2) > \sqrt{(2k_1 - k_3^2)^2 + 4k_2^2}$. Take the helix $\gamma_{\omega,\sigma,h}$ of type 1 determined by

$$\omega^2 = \frac{1}{2}\left(2k_1 + k_3^2\right) - \frac{1}{2}\sqrt{(2k_1 - k_3^2)^2 + 4k_2^2},$$
$$\sigma^2 = \frac{1}{2}\left(2k_1 + k_3^2\right) + \frac{1}{2}\sqrt{(2k_1 - k_3^2)^2 + 4k_2^2},$$
$$h^2 = \frac{k_3^2}{2k_1k_3^2 - k_2^2}.$$

A straightforward computation shows that $\frac{1}{\sigma^2} < h^2 < \frac{1}{\omega^2}$ and that the curvatures of $\gamma_{\omega,\sigma,h}$ are k_1, k_2 and k_3.

Case 2: Suppose that $(2k_1 + k_3^2) < \sqrt{(2k_1 - k_3^2)^2 + 4k_2^2}$. Take the helix $\gamma_{\omega,\sigma,h}$ of type 2 determined by

$$\omega^2 = -\frac{1}{2}\left(2k_1 + k_3^2\right) + \frac{1}{2}\sqrt{(2k_1 - k_3^2)^2 + 4k_2^2},$$
$$\sigma^2 = \frac{1}{2}\left(2k_1 + k_3^2\right) + \frac{1}{2}\sqrt{(2k_1 - k_3^2)^2 + 4k_2^2},$$
$$h^2 = \frac{k_3^2}{2k_1k_3^2 - k_2^2}.$$

As before we have that the curvatures of $\gamma_{\omega,\sigma,h}$ are k_1, k_2 and k_3.

Case 3: Finally, if $(2k_1 + k_3^2) = \sqrt{(2k_1 - k_3^2)^2 + 4k_2^2}$, then we can take the helix $\gamma_{\sigma,h}$ determined by

$$\sigma^2 = \frac{k_2^2 + k_3^4}{k_3^2} \quad \text{and} \quad h^2 = \frac{k_3^4}{2\left(k_2^2 + k_3^4\right)}.$$

It is easy to see that $h^2 < 1/2$ and the curvatures of $\gamma_{\sigma,h}$ are k_1, k_2 and k_3. That concludes the proof.

□

7.4 Lightlike Curves in $\mathbb{S}_1^n$

Let $\gamma : I \to \mathbb{S}_1^n \subset \mathbb{R}_1^{n+1}$ be a lightlike curve and denote by D_t the covariant derivative in $\mathbb{S}_1^n$ along γ. Then for any vector field V along γ we have

$$\frac{DV}{dt} = V' + \langle V, \gamma' \rangle \gamma,$$

where $\langle , \rangle$ stands for the canonical metric in $\mathbb{R}_1^{n+1}$. If $\{L, N, W_1, \dots, W_m\}$ denotes the Cartan frame, then Eq. (7.4) write down as follows:

$$\begin{aligned}
\gamma' &= L,\\
L' &= W_1,\\
N' &= -\gamma + k_1 W_1 + k_2 W_2,\\
W_1' &= -k_1 L - N,\\
W_2' &= -k_2 L + k_3 W_3,\\
W_i' &= -k_i W_{i-1} + k_{i+1} W_{i+1}, \quad i \in \{3, \dots, m-1\},\\
W_m' &= -k_m W_{m-1}.
\end{aligned} \tag{7.6}$$

Theorem 7.4.1 *Let $k_1, k_2, \dots, k_m : [-\varepsilon, \varepsilon] \to \mathbb{R}$ be differentiable functions with $k_2 < 0$ and $k_i > 0$ for $i \in \{3, \dots, m-1\}$. Let p be a point in $\mathbb{S}_1^n$, $n = m + 2$, and let $\{L^0, N^0, W_1^0, \dots, W_m^0\}$ be a positively oriented pseudo-orthonormal basis of $T_p\mathbb{S}_1^n$. Then there exists a unique Cartan curve γ in $\mathbb{S}_1^n$, with $\gamma(0) = p$, whose Cartan frame $\{L, N, W_1, \dots, W_m\}$ satisfies*

$$L(0) = L^0, N(0) = N^0, W_i(0) = W_i^0, \quad i \in \{1, \dots, m\}.$$

Proof We proceed as in the proof of Theorem 7.3.3. Let $\{L(t), N(t), \gamma(t), W_1(t), \dots, W_m(t)\}$ be a solution of (7.6) with initial conditions $\{L^0, N^0, p, W_1^0, \dots, W_m^0\}$. A straightforward computation yields

$$\frac{d}{dt}\left(L_i(t)N_j(t) + L_j(t)N_i(t) + \gamma_i(t)\gamma_j(t) + \sum_{\beta=1}^{m} W_{\beta i}(t) W_{\beta j}(t) \right) = 0.$$

Since $\{L^0, N^0, p, W_1^0, \dots, W_m^0\}$ is a pseudo-orthonormal basis, from Lemma 7.3.2, with $q = r = 1$, we obtain

$$L_i(t)N_j(t) + L_j(t)N_i(t) + \alpha_i(t)\gamma_j(t) + \sum_{\beta=1}^{m} W_{\beta i}(t) W_{\beta j}(t) = \eta_{ij}, \quad \forall t \in [-\varepsilon, \varepsilon],$$

and then $\{L(t), N(t), \gamma(t), W_1(t), \dots, W_m(t)\}$ is a local pseudo-orthonormal frame. This concludes the proof. □

The following result is clear.

Theorem 7.4.2 *If two Cartan curves C and $\bar{C}$ in $\mathbb{S}_1^n$ have Cartan curvatures $\{k_1, \dots, k_n\}$, where $k_i : [-\varepsilon, \varepsilon] \to \mathbb{R}$ are differentiable functions, then there exists a Lorentzian transformation of $\mathbb{S}_1^n$ which maps C into $\bar{C}$.*

7.4.1 Lightlike Helices in $\mathbb{S}_1^4$

In this section we are going to classify the lightlike helices in the 4-dimensional De Sitter space. The Cartan frame of a lightlike curve γ in $\mathbb{S}_1^4$ satisfies the following equations

$$\begin{aligned} \gamma' &= L, \\ L' &= W_1, \\ N' &= -\gamma + k_1 W_1 + k_2 W_2, \\ W_1' &= -k_1 L - N, \\ W_2' &= -k_2 L. \end{aligned} \tag{7.7}$$

An easy computation shows that if γ is a helix, then it satisfies the following differential equation

$$\gamma^{(5)} + 2k_1\gamma^{(3)} - (1 + k_2^2)\gamma' = 0,$$

whose general solution is

$$\gamma(t) = A_1 \sinh \omega t + A_2 \cosh \omega t + A_3 \sin \sigma t + A_4 \cos \sigma t + A_5,$$

where A_1, A_2, A_3, A_4 and A_5 are constant vectors in $\mathbb{R}_1^5$.

Example 7.4.3 Let $\gamma_{\omega,\sigma} : \mathbb{R} \to \mathbb{S}_1^4 \subset \mathbb{R}_1^5$ be the lightlike curve defined by

$$\gamma_{\omega,\sigma}(t) = \sqrt{\frac{1}{\omega^2 + \sigma^2}} \times \left(\frac{1}{\omega} \sinh \omega t, \frac{1}{\omega} \cosh \omega t, \frac{1}{\sigma} \sin \sigma t, \frac{1}{\sigma} \cos \sigma t, \sqrt{\frac{\omega^4 - 1}{\omega^2} + \frac{\sigma^4 - 1}{\sigma^2}} \right),$$

where ω and σ are non-zero constants such that $\omega^2\sigma^2 > 1$.

A direct computation shows that γ is a lightlike helix with curvatures

$$k_1 = \frac{1}{2}\left(\sigma^2 - \omega^2\right) \quad \text{and} \quad k_2^2 = \omega^2\sigma^2 - 1. \tag{7.8}$$

Theorem 7.4.4 *Let γ be a lightlike curve fully immersed in $\mathbb{S}_1^4$. Then γ is a helix if and only if it is congruent to one of the family described in Example 7.4.3.*

Proof Let γ be a lightlike helix in $\mathbb{S}_1^4$ with curvatures k_1 and k_2. Consider the constants ω and σ given by

$$\omega = \sqrt{-k_1 + \sqrt{k_1^2 + k_2^2 + 1}} \quad \text{and} \quad \sigma = \sqrt{k_1 + \sqrt{k_1^2 + k_2^2 + 1}}.$$

Then from (7.8) we deduce that $\gamma_{\omega,\sigma}$ has curvatures k_1 and k_2, so that we get the result by using the congruence Theorem 7.4.2. □

7.5 Lightlike Curves in $\mathbb{H}_1^n$

Let $\gamma : I \to \mathbb{H}_1^n \subset \mathbb{R}_2^{n+1}$ be a lightlike curve and denote by D_t the covariant derivative in $\mathbb{H}_1^n$ along γ. Then for any vector field V along γ we have

$$\frac{DV}{dt} = V' - \langle V, \gamma' \rangle \gamma,$$

where $\langle\, ,\, \rangle$ stands for the canonical metric in $\mathbb{R}_2^{n+1}$. If $\{L, N, W_1, \ldots, W_m\}$ denotes the Cartan frame, then we have the following equations

$$\begin{aligned}
\gamma' &= L, \\
L' &= W_1, \\
N' &= \gamma + k_1 W_1 + k_2 W_2, \\
W_1' &= -k_1 L - N, \\
W_2' &= -k_2 L + k_3 W_3, \\
W_i' &= -k_i W_{i-1} + k_{i+1} W_{i+1}, \quad i \in \{3, \ldots, m-1\}, \\
W_m' &= -k_m W_{m-1}.
\end{aligned} \tag{7.9}$$

Theorem 7.5.1 *Let $k_1, k_2, \ldots, k_m : [-\varepsilon, \varepsilon] \to \mathbb{R}$ be differentiable functions with $k_2 < 0$ and $k_i > 0$ for $i \in \{3, \ldots, m-1\}$. Let p be a point in $\mathbb{H}_1^n$, $n = m + 2$, and let $\{L^0, N^0, W_1^0, \ldots, W_m^0\}$ be a positively oriented pseudo-orthonormal basis of $T_p\mathbb{H}_1^n$. Then there exists a unique Cartan curve γ in $\mathbb{H}_1^n$, with $\gamma(0) = p$, whose*

Cartan frame $\{L, N, W_1, \ldots, W_m\}$ *satisfies*

$$L(0) = L^0, N(0) = N^0, W_i(0) = W_i^0, \quad i \in \{1, \ldots, m\}.$$

Proof We proceed as in the proof of Theorem 7.4.1. Let $\{L(t), N(t), \gamma(t), W_1(t), \ldots, W_m(t)\}$ be a solution of (7.9) with initial conditions $\{L^0, N^0, p, W_1^0, \ldots, W_m^0\}$. A straightforward computation yields

$$\frac{d}{dt}\left(L_i(t)N_j(t) + L_j(t)N_i(t) - \gamma_i(t)\alpha_j(t) + \sum_{\beta=1}^{m} W_{\beta i}(t)W_{\beta j}(t)\right) = 0.$$

Since $\{L^0, N^0, p, W_1^0, \ldots, W_m^0\}$ is a pseudo-orthonormal basis, from Lemma 7.3.2, with $q = 2$ and $r = 1$, we obtain

$$L_i(t)N_j(t) + L_j(t)N_i(t) - \gamma_i(t)\gamma_j(t) + \sum_{\beta=1}^{m} W_{\beta i}(t)W_{\beta j}(t) = \eta_{ij}, \quad \forall t \in [-\varepsilon, \varepsilon],$$

and then $\{L(t), N(t), \gamma(t), W_1(t), \ldots, W_m(t)\}$ is a local pseudo-orthonormal frame with $\langle \gamma, \gamma \rangle = -1$. That concludes the proof. □

The following result is clear.

Theorem 7.5.2 *If two Cartan curves C and $\bar{C}$ in $\mathbb{H}_1^n$ have Cartan curvatures $\{k_1, \ldots, k_n\}$, where $k_i : [-\varepsilon, \varepsilon] \to \mathbb{R}$ are differentiable functions, then there exists a Lorentzian transformation of $\mathbb{H}_1^n$ which maps C into $\bar{C}$.*

7.5.1 Lightlike Helices in $\mathbb{H}_1^4$

Let γ be a lightlike curve in $\mathbb{H}_1^4$, then its Cartan frame satisfies the following equations

$$\begin{aligned} \gamma' &= L, \\ L' &= W_1, \\ N' &= \gamma + k_1 W_1 + k_2 W_2, \\ W_1' &= -k_1 L - N, \\ W_2' &= -k_2 L. \end{aligned} \tag{7.10}$$

If γ is a helix, then it is easy to see that it verifies the following ordinary differential equation

$$\gamma^{(5)} + 2k_1\gamma^{(3)} + (1 - k_2^2)\gamma' = 0.$$

Before we state the main result of this section we present some examples of helices in the 4-dimensional anti De Sitter space.

Example 7.5.3 (Helices of Type A1) Let $0 < \omega^2 < 1$ and let γ_ω be the curve in $\mathbb{H}_1^4$ defined by

$$\gamma_\omega(t) = \left(\frac{t}{2\omega}\cosh\omega t, \frac{1}{\omega^2}\left(\cosh\omega t - \frac{1}{2}\omega t\sinh\omega t\right), \frac{1}{\omega^2}\left(\sinh\omega t - \frac{1}{2}\omega t\cosh\omega t\right),\right.$$
$$\left.\frac{t}{2\omega}\sinh\omega t, \frac{\sqrt{1-\omega^4}}{\omega^2}\right).$$

Then γ_ω is a helix with curvatures

$$k_1 = -\omega^2 \quad \text{and} \quad k_2^2 = 1 - \omega^4.$$

Example 7.5.4 (Helices of Type A2) Let $0 < \sigma^2 < 1$ and let γ_σ be the curve in $\mathbb{H}_1^4$ defined by

$$\gamma_\sigma(t) = \left(\frac{1}{\sigma^2}\left(\sin\sigma t - \frac{1}{2}\sigma t\cos\sigma t\right), \frac{1}{\sigma^2}\left(\cos\sigma t + \frac{1}{2}\sigma t\sin\sigma t\right), -\frac{t}{2\sigma}\cos\sigma t,\right.$$
$$\left.\frac{t}{2\sigma}\sin\sigma t, \frac{\sqrt{1-\sigma^4}}{\sigma^2}\right).$$

Then γ_σ is a helix with curvatures

$$k_1 = \sigma^2 \quad \text{and} \quad k_2^2 = 1 - \sigma^4.$$

Example 7.5.5 (Helices of Type A3) Let $\omega^2 = 1$ and let γ be the lightlike curve in $\mathbb{H}_1^4$ defined by

$$\gamma(t) = \left(1 - \frac{t^4}{24}, \frac{\omega(t^3+t)}{2\sqrt{3}}, \frac{t^4}{24}, \frac{\omega(t^3-t)}{2\sqrt{3}}, \frac{t^2}{2}\right).$$

Then γ is a helix with curvatures

$$k_1 = 0 \quad \text{and} \quad k_2^2 = 1.$$

α will be called the *lightlike quartic* in $\mathbb{H}_1^4$.

Example 7.5.6 (Helices of Type B1) Let $0 < \omega^2 < \sigma^2$ and $\omega^2\sigma^2 < 1$, and let $\gamma_{\omega,\sigma}$ be the curve in $\mathbb{H}_1^4$ defined by

$$\gamma_{\omega,\sigma}(t) = \sqrt{\frac{1}{\sigma^2-\omega^2}}\left(\frac{1}{\omega}\sin\omega t, \frac{1}{\omega}\cos\omega t, \frac{1}{\sigma}\sin\sigma t, \frac{1}{\sigma}\cos\sigma t, \sqrt{\frac{1+\omega^4}{\omega^2}-\frac{1+\sigma^4}{\sigma^2}}\right).$$

Then $\gamma_{\omega,\sigma}$ is a helix with curvatures

$$k_1 = \frac{1}{2}\left(\omega^2+\sigma^2\right) \quad \text{and} \quad k_2^2 = 1-\omega^2\sigma^2.$$

Example 7.5.7 (Helices of Type B2) Let $0 < \sigma^2 < \omega^2$ and $\omega^2\sigma^2 < 1$, and let$\gamma_{\omega,\sigma}$ be the curve in $\mathbb{H}_1^4$ defined by

$$\gamma_{\omega,\sigma}(t) = \sqrt{\frac{1}{\omega^2-\sigma^2}}$$
$$\times\left(\frac{1}{\omega}\sinh\omega t, \frac{1}{\sigma}\cosh\sigma t, \frac{1}{\sigma}\sinh\sigma t, \frac{1}{\omega}\cosh\omega t, \sqrt{\frac{1+\sigma^4}{\sigma^2}-\frac{1+\omega^4}{\omega^2}}\right).$$

Then $\gamma_{\omega,\sigma}$ is a helix with curvatures

$$k_1 = -\frac{1}{2}\left(\omega^2+\sigma^2\right) \quad \text{and} \quad k_2^2 = 1-\omega^2\sigma^2.$$

Example 7.5.8 (Helices of Type B3) Let $\sigma \neq 0$ and let γ_σ be the curve in $\mathbb{H}_1^4$ defined by

$$\gamma_\sigma(t) = \left(\frac{2+2\sigma^4-\sigma^2t^2}{2\sigma^2\sqrt{1+\sigma^4}}, \frac{t}{\sigma}, \frac{t^2}{2\sqrt{1+\sigma^4}}, \frac{1}{\sigma^2}\sin\sigma t, \frac{1}{\sigma^2}\cos\sigma t\right).$$

Then γ_σ is a helix with curvatures

$$k_1 = \frac{\sigma^2}{2} \quad \text{and} \quad k_2^2 = 1.$$

Example 7.5.9 (Helices of Type B4) Let $\omega \neq 0$ and let γ_ω be the curve in $\mathbb{H}_1^4$ defined by

$$\gamma_\omega(t) = \left(\frac{2+2\omega^4+\omega^2t^2}{2\omega^2\sqrt{1+\omega^4}}, \frac{1}{\omega^2}\sinh\omega t, \frac{1}{\omega^2}\cosh\omega t, \frac{t^2}{2\sqrt{1+\omega^4}}, \frac{t}{\omega}\right).$$

Then γ_ω is a helix with curvatures

$$k_1 = -\frac{\omega^2}{2} \quad \text{and} \quad k_2^2 = 1.$$

Example 7.5.10 (Helices of Type B5) Let $\omega\sigma \neq 0$ and let $\gamma_{\omega,\sigma}$ be the curve in $\mathbb{H}_1^4$ defined by

$$\gamma_{\omega,\sigma}(t) = \sqrt{\frac{1}{\omega^2+\sigma^2}} \times \left(\sqrt{\frac{1+\omega^4}{\omega^2} + \frac{1+\sigma^4}{\sigma^2}}, \frac{1}{\omega}\sinh \omega t, \frac{1}{\omega}\cosh \omega t, \frac{1}{\sigma}\sin \sigma t, \frac{1}{\sigma}\cos \sigma t \right).$$

Then $\gamma_{\omega,\sigma}$ is a helix with curvatures

$$k_1 = \frac{1}{2}\left(\sigma^2 - \omega^2\right) \quad \text{and} \quad k_2^2 = 1 + \omega^2\sigma^2.$$

Example 7.5.11 (Helices of Type C) Let $\omega^2 + \sigma^2 < 1$ and let $\gamma_{\omega,\sigma}$ be the curve in $\mathbb{H}_1^4$ defined by

$$\gamma_{\omega,\sigma}(t) = \frac{1}{2\omega\sigma\left(\omega^2+\sigma^2\right)} \Big(2\omega\sigma \cosh \omega t \sin \sigma t + \left(\omega^2 - \sigma^2\right) \sinh \omega t \cos \sigma t,$$
$$-2\omega\sigma \cosh \omega t \cos \sigma t + \left(\omega^2 - \sigma^2\right) \sinh \omega t \sin \sigma t, \left(\omega^2 + \sigma^2\right) \sinh \omega t \cos \sigma t,$$
$$\left(\omega^2 + \sigma^2\right) \sinh \omega t \sin \sigma t, 2\omega\sigma\sqrt{1 - \left(\omega^2+\sigma^2\right)^2} \Big).$$

Then $\gamma_{\omega,\sigma}$ is a helix with curvatures

$$k_1 = -\omega^2 + \sigma^2 \quad \text{and} \quad k_2^2 = 1 - \left(\omega^2 + \sigma^2\right)^2.$$

Theorem 7.5.12 *Let γ be a lightlike curve fully immersed in $\mathbb{H}_1^4$. Then γ is a helix if and only if it is congruent to one helix of the families described in Examples 7.5.3–7.5.11.*

Proof That result can be obtained following a similar reasoning as in the precedent cases. The idea of the proof consists of finding, for any constants k_1 and $k_2 \neq 0$, a helix in one of the above nine examples with curvatures k_1 and k_2. The following table collects all the possibilities and Fig. 7.1 represents a diagram of them. □

Case	Type	Parameters
$k_1^2 + k_2^2 = 1, k_1 < 0, 0 < k_2^2 < 1$	A1	$\omega^2 = -k_1$
$k_1^2 + k_2^2 = 1, k_1 > 0, 0 < k_2^2 < 1$	A2	$\sigma^2 = k_1$
$k_1 = 0, k_2^2 = 1$	A3	
$k_1^2 + k_2^2 > 1, k_1 > 0, 0 < k_2^2 < 1$ $\sigma^2 = k_1 + \sqrt{k_1^2 + k_2^2 - 1}$	B1	$\omega^2 = k_1 - \sqrt{k_1^2 + k_2^2 - 1},$
$k_1^2 + k_2^2 > 1, k_1 < 0, 0 < k_2^2 < 1$ $\sigma^2 = -k_1 - \sqrt{k_1^2 + k_2^2 - 1}$	B2	$\omega^2 = -k_1 + \sqrt{k_1^2 + k_2^2 - 1},$
$k_1^2 + k_2^2 > 1, k_1 > 0, k_2^2 = 1$	B3	$\sigma^2 = 2k_1$
$k_1^2 + k_2^2 > 1, k_1 < 0, k_2^2 = 1$	B4	$\omega^2 = -2k_1$
$k_2^2 > 1$ $\sigma^2 = k_1 + \sqrt{k_1^2 + k_2^2 - 1}$	B5	$\omega^2 = -k_1 + \sqrt{k_1^2 + k_2^2 - 1},$
$k_1^2 + k_2^2 < 1$ $\sigma^2 = \frac{1}{2}\left(k_1 + \sqrt{1 - k_2^2}\right)$	C	$\omega^2 = \frac{1}{2}\left(-k_1 + \sqrt{1 - k_2^2}\right),$

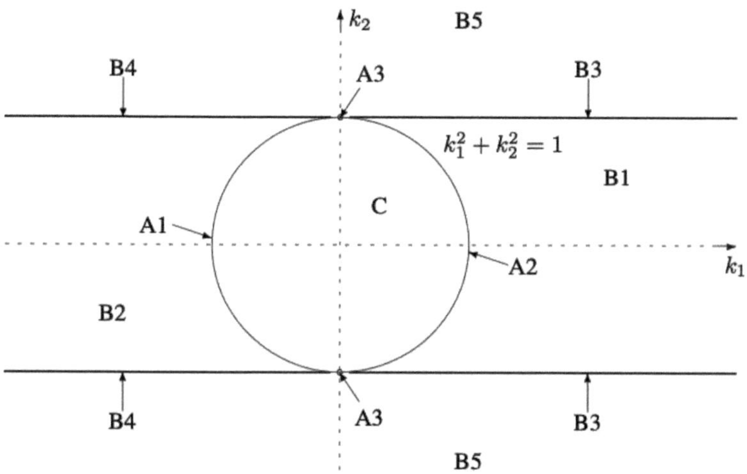

Fig. 7.1 Regions in $\mathbb{R}^2$ with its corresponding types of helices

Chapter 8
Lightlike Lancret Helices in $\mathbb{L}^n$

Remember that a generalized, or Lancret, helix in $\mathbb{R}^3$ is a curve of constant slope, in other words, a curve whose tangent makes a constant angle with a fixed direction (called the axis). In addition, a curve whose tangent indicatrix is a plane curve. The study of such a curves in $\mathbb{R}^3$ dates from 1802 when M. A. Lancret stated that (see [94] for details) "a curve is a Lancret helix if and only if the ratio of curvature to torsion is constant".

The n-dimensional case ($n = 2m + 1$ odd) was considered by Hayden in 1931 (see [55]), who called a "Lancret helix" a curve satisfying that the ratios $\kappa_{2i}/\kappa_{2i-1}$, $i = 1, \ldots, m$, are constant, $\kappa_1, \kappa_2, \ldots, \kappa_{n-1}$ being the curvatures of the curve. Hayden proved in [55, 56] that a curve is a Lancret helix if there exists a parallel vector field lying in the osculating space of the curve which makes constant angles with the tangent and the principal normals.

The Lancret theorem was revisited and solved by M. Barros (see [6]) in 3-dimensional real space forms by using Killing vector fields along curves. Recently, new improvements have been achieved in Lorentzian space forms. In [11], a non-null curve γ immersed in $\mathbb{L}^3$ is called a Lancret helix if its tangent indicatrix is laid in a plane $\pi \subset \mathbb{L}^3$. Then according to the causal character of π the authors have to distinguish between degenerate and non-degenerate Lancret helices and show the corresponding Lancret theorem.

To point out the interest of non-null Lancret helices it should be mentioned that they arise in the context of the interplay between geometry and integrable Hamiltonian systems (see [67] and [68]). In [10] we have found parameterized solutions of the localized induction equation (LIE) in the 3-dimensional anti De Sitter space, so that the soliton solutions are the lightlike geodesics of the Lorentzian Hopf cylinders. Therefore there is a natural geometric evolution on Lancret helices inducing a modified Korteweg-de Vries curvature evolution equation coming from LIE. The role of Lancret helices here is probably similar to that of curves of constant torsion or constant natural curvature (see [68]).

A. Ferrandez Izquierdo, *Helices*, RSME Springer Series 16,
https://doi.org/10.1007/978-3-032-09663-0_8

Other applications of Lancret helices can be found in [71], where the author proposed a mathematical model of the auditory process in the cochlea that does not neglect the effect of cochlea coiling, and [46], where the authors obtained, for inhomogeneous electromagnetic waves in isotropic media, the operator evolution solutions of Maxwell equations; in the case of homogeneous waves an evolution operator is associated with a set of right-handed and left-handed Lancret helices.

It is well known that in the geometry of lightlike curves the natural parameter is the pseudo-arc (see [24] and [104]). In [43] we generalize the results of Bonnor by introducing a Frenet frame (which we call the Cartan frame) along a lightlike curve with the minimun number of curvature functions and classify lightlike helices. Now we use the Cartan frame (and the Cartan curvatures) introduced there to define and study lightlike Lancret helices in $\mathbb{L}^n$ (some results for lightlike curves in 3-dimensional spaces are obtained in [5]).

This chapter is taken from [44].

8.1 Frenet Equations

We follow the notation and terminology stated in [43]. Let M_1^n be an orientable Lorentzian manifold and let $\gamma : I \subset \mathbb{R} \longrightarrow M_1^n$ be a lightlike curve parameterized by the pseudo-arc. We have the following result.

Theorem 8.1.1 ([43]) *Assume that* $\left\{\gamma'(t), \gamma''(t), \ldots, \gamma^{(n)}(t)\right\}$ *is a basis of* $T_{\gamma(t)} M_1^n$ *for all* t. *Then there exists exactly one Frenet frame* $\{L, W_1, N, W_2, \ldots, W_m\}$ $(m = n-2)$ *satisfying the equations*

$$\begin{aligned} L' &= W_1, \\ W_1' &= -k_1 L + N, \\ N' &= -k_1 W_1 + k_2 W_2, \\ W_2' &= k_2 L + k_3 W_3, \\ W_i' &= -k_i W_{i-1} + k_{i+1} W_{i+1} \quad i \in \{3, \ldots, m-1\}, \\ W_m' &= -k_m W_{m-1}, \end{aligned} \tag{8.1}$$

and fulfilling

(i) $\left\{\gamma', \gamma'', \ldots, \gamma^{(i)}\right\}$ *and* $\{L, W_1, N, W_2, \ldots, W_{i-2}\}$ *have the same orientation for* $1 \leqslant i \leqslant m-1$,

(ii) $\{L, W_1, N, W_2, \ldots, W_m\}$ *is positively oriented.*

Furthermore, the curvature functions satisfy $k_i > 0$ *for all* $i \geq 2$.

The above Frenet frame and curvature functions are called the *Cartan frame* and the *Cartan curvatures* of the lightlike curve γ.

8.1.1 Frenet Frames for Spacelike Curves in Lightlike Totally Geodesic Submanifolds of a Lorentzian Space Form

Let N^m be a lightlike totally geodesic submanifold in an oriented Lorentzian manifold M_1^n and let $\beta : J \subset \mathbb{R} \longrightarrow N^m$ be a locally parameterized spacelike curve. Let us assume that $\left\{\beta'(s), \ldots, \beta^{(m)}(s)\right\}$ is a basis of $T_{\beta(s)}N^m$, where s stands for the arc length parameter. Write $E_i(s) = \text{span}\left\{\beta'(s), \ldots, \beta^{(i)}(s)\right\}$, for $1 \leq i \leq m$, and assume that $\dim\{\text{rad}(E_i(s))\}$ is constant for all $s \in J$. Since β is contained in a lightlike submanifold, there exist an index $1 \leq i_0 \leq m$ such that $\dim\ \text{rad}(E_{i_0}) = 1$. Let us denote $r = \min\{i : \dim\ \text{rad}(E_i) = 1\}$, then $r > 1$ because β is spacelike, and $\dim\ \text{rad}(E_j) = 1$ for $j > r$.

Now we are going to construct a Frenet frame for this kind of curves. The first vector will be $T(s) = \beta'(s)$. As in the non-degenerate case, and using the Gram-Schmidt method applied to E_{r-1}, we can construct a set of orthonormal spacelike vectors $\{T, V_1, \ldots, V_{r-2}\}$ such that $E_{j+1} = \text{span}\left\{T, V_1, \ldots, V_j\right\}$, for $1 \leq j \leq r-2$. Since $\dim\{\text{rad}(E_r)\} = 1$, we can find a vector $\mathcal{L}$ (not unique) such that

$$E_r = E_{r-1} \perp \text{span}\{\mathcal{L}\} = \text{span}\{T, V_1, \ldots, V_{r-2}\} \perp \text{span}\{\mathcal{L}\},$$

where $\perp$ denotes the direct orthogonal sum. A straightforward computation shows that the following equations hold:

$$\begin{aligned} \beta'(s) &= T \\ T' &= \rho_1 V_1 \\ V_1' &= -\rho_1 T + \rho_2 V_2 \\ V_i' &= -\rho_i V_{i-1} + \rho_{i+1} V_{i+1} \quad 2 \leq i \leq r-3 \\ V_{r-2}' &= -\rho_{r-2} V_{r-3} + \rho_{r-1} \mathcal{L} \end{aligned} \tag{8.2}$$

where $\rho_j : J \longrightarrow \mathbb{R}$ are differentiable functions and $(\,)'$ denotes covariant derivative in N^m, which in this case agrees with the covariant derivative of the ambient space M_1^n.

Now it is easy to see $r = m$, so that the Frenet frame is $\{T, V_1, \ldots, V_{m-2}, \mathcal{L}\}$. That frame, as in the non-degenerate case, can be constructed in a unique way (up to orientation) except $\mathcal{L}$. The vector $\mathcal{L}$ can be arbitrarily chosen depending on each situation; a good choice is $\rho_{m-1} = \pm 1$. In any case we need at least $m-1$ curvature functions in order to determine completely the curve. Moreover, the most natural criterion to choose the orientation is to consider that $\left\{\beta', \ldots, \beta^{(i+1)}\right\}$ and $\{T, V_1, \ldots, V_i\}$, $1 \leq i \leq m-2$ have the same orientation, and that $\{T, V_1, \ldots, V_{m-2}, \mathcal{L}\}$ is positively oriented.

The following theorems of existence, uniqueness and congruence can be proved in a similar way as in [43]. Now $M_1^n(c)$ denotes a Lorentzian space form of constant curvature c.

Theorem 8.1.2 *Let $\rho_1, \rho_2, \ldots, \rho_m : [-\varepsilon, \varepsilon] \longrightarrow \mathbb{R}$ be differentiable functions. Let N^m be a lightlike totally geodesic submanifold of $M_1^n(c)$, p a point in N^m and consider a positively oriented pseudo-orthonormal basis $\{T^0, V_1^0, \ldots, V_{m-2}^0, \mathcal{L}^0\}$ of T_pN^m. Then there exists a unique spacelike Cartan curve α in $M_1^n(c)$, contained in N^m with $\alpha(0) = p$, whose Cartan frame $\{T, V_1, \ldots, V_{m-2}, \mathcal{L}\}$ satisfies*

$$T(0) = T^0, V_1(0) = V_1^0, \ldots, V_{m-2}(0) = V_{m-2}^0, \mathcal{L}(0) = \mathcal{L}^0.$$

Theorem 8.1.3 *Let C and $\overline{C}$ be two spacelike Cartan curves laid in a lightlike totally geodesic submanifold $N^m \subset M_1^n(c)$, having Cartan curvatures $\{\rho_1, \ldots, \rho_m\}$ in N^m. Then there exists a Lorentzian transformation of M_1^n (or N^m) which maps C into $\overline{C}$.*

8.2 Lightlike Lancret Helices in $\mathbb{L}^n$

Let $\gamma : I \longrightarrow \mathbb{L}^n$, $n = 2q + 3$, be a locally parameterized lightlike Cartan curve and set $m = n - 2 = 2q + 1$. In a similar way to the non-degenerate case (see [55, 56, 96, 97, 107, 108]), we present the following definition.

Definition 8.2.1 A lightlike Cartan curve $\gamma : I \longrightarrow \mathbb{L}^n$, $n = 2q + 3$, is said to be a *Lancret helix* if there exist a non-zero constant vector v such that all the products $\langle L(t), v\rangle \neq 0$, $\langle N(t), v\rangle$ and $\langle W_{2i+1}(t), v\rangle \neq 0$, $1 \leq i \leq q - 1$, are constant.

In the non-degenerate case, the vectors appearing in the definition are unitary and the constancy of the products implies that the curve γ makes a constant angle with some of the vectors of the Cartan frame. The straight line generated by v, which can be spacelike, timelike or lightlike, is uniquely determined and will be called the *axis* of γ. When v is a non-lightlike vector (i.e. spacelike or timelike), we can assume without loss of generality that v is unitary.

A classical result stated by M. A. Lancret in 1802 and first proved by B. de Saint Venant in 1845 (see [94] for details) says that "a curve in $\mathbb{R}^3$ is a Lancret helix if and only if the ratio of curvature to torsion is constant". A straightforward computation from Theorem 8.1.1 yields to a generalization of this result for lightlike curves in $\mathbb{L}^n$.

Theorem 8.2.2 (The Lancret Theorem for lightlike Curves) *Let $\gamma : I \longrightarrow \mathbb{L}^n$ be a lightlike Cartan curve. Then the following statements are equivalent:*

(i) *There exist constants $\{r, r_1, \ldots, r_q\}$ ($r_i \neq 0$) such that*

$$k_1(t) = r \quad \textit{and} \quad k_{2i+1}(t) = r_i k_{2i}(t) \quad \textit{for } 1 \leq i \leq q, \quad n = 2q + 3.$$

(ii) *γ is a Lancret helix.*

Proof (i) $\Rightarrow$ (ii). Let us denote $R_i = \prod_{j \geq i} r_j$ and consider the vector field along the curve γ given by

$$v(t) = R_1(rL + N) + \sum_{i=1}^{q-1} R_{i+1} W_{2i+1} + W_m.$$

Taking into account (8.1), statement (i) and that $r_i R_{i+1} = R_i$, we have that $v(t)$ is a constant vector. Moreover, it is easy to see that $\langle v, L \rangle$, $\langle v, N \rangle$ and $\langle v, W_{2i+1} \rangle$ are constant, and that concludes the first part of the proof.

(ii) $\Rightarrow$ (i). Let v be the axis. Then it is not difficult to see that v is orthogonal to W_{2i+2}, $1 \leq i \leq q-1$, so we can write

$$v = \varrho L + \varrho_1 N + \sum_{i=1}^{q} \varrho_{2i+1} W_{2i+1},$$

for certain functions ϱ and ϱ_j. From the hypothesis we easily deduce that ϱ and ϱ_j are constant, for $j \leq m$. Finally, from $v'(t) = 0$ we get

$$k_1 = \frac{\varrho}{\varrho_1} = r, \quad k_{2i+1} = \frac{\varrho_{2i-1}}{\varrho_{2i+1}} k_{2i} = r_i k_{2i},$$

and the proof finishes. □

8.3 Lightlike Lancret Helices in Low Dimensions

8.3.1 The 3-Dimensional Case

A lightlike Cartan curve $\gamma : I \longrightarrow \mathbb{L}^3$ is a Lancret helix if there exist a non-zero constant vector v such that $\langle \gamma', v \rangle$ is constant. That means that the tangent indicatrix is laid in a plane, or likewise, there exists a non-zero vector v in $\mathbb{L}^3$ which is orthogonal to the acceleration vector field of γ. The following result is an easy consequence from Theorem 8.2.2.

Proposition 8.3.1 *Let $\gamma : I \longrightarrow \mathbb{L}^3$ be a lightlike Cartan curve. Then γ is a Lancret helix if and only if γ is a Cartan helix.*

It is well-known that, up to congruences, there are exactly three types of helices, according to its curvature function (or its axis):

Curve	Curvature	Axis
$\gamma(t) = \left(-\frac{t}{\sigma}, \frac{1}{\sigma^2}\sin\sigma t, \frac{1}{\sigma^2}\cos\sigma t\right)$	$k = \frac{1}{2}\sigma^2 > 0$	$\upsilon = (1, 0, 0)$ timelike
$\gamma(t) = \left(\frac{1}{\omega^2}\sinh\omega t, \frac{1}{\omega^2}\cosh\omega t, -\frac{t}{\omega}\right)$	$k = -\frac{1}{2}\omega^2 < 0$	$\upsilon = (0, 0, 1)$ spacelike
$\gamma(t) = \left(\frac{t^3}{4} + \frac{t}{3}, \frac{t^2}{2}, \frac{t^3}{4} - \frac{t}{3}\right)$	$k = 0$	$\upsilon = (1, 0, 1)$lightlike.

8.3.2 The 5-Dimensional Case with Non-lightlike Axis

A lightlike Cartan curve Γ in $\mathbb{L}^5$ is a Lancret helix if there exist a non-zero constant vector υ satisfying that $\langle \upsilon, L\rangle$ and $\langle \upsilon, N\rangle$ are constant. Let υ be a unit vector and set $\langle \upsilon, \upsilon\rangle = \varepsilon$. Let Σ denotes the hyperplane orthogonal to υ, P the projection map onto the hyperplane Σ and $\Upsilon = P(\Gamma)$ the projected curve. Let $\gamma : I \longrightarrow \mathbb{L}^5$ be the pseudo-arc parameterization of Γ and let $\overline{\beta} : I \longrightarrow \Sigma$ be the parameterization of Υ with respect to the same parameter t. Then we can write

$$\gamma(t) = \overline{\beta}(t) + \overline{\mu}(t)\upsilon, \tag{8.3}$$

where $\overline{\mu} : I \longrightarrow \mathbb{R}$ is a non-constant differentiable function. By differentiating we have

$$L(t) = \overline{\beta}'(t) + \overline{\mu}'(t)\upsilon \qquad \text{and} \qquad \left\langle \overline{\beta}'(t), \overline{\beta}'(t)\right\rangle = -\varepsilon\,\overline{\mu}'(t)^2.$$

Then the projected curve Υ is spacelike (resp. timelike) according to υ is timelike (resp. spacelike). Let $\beta : J \longrightarrow \Sigma$ be the arc-length parametrization of Υ and let s denotes the arc-length parameter. Let us consider its Frenet reference $\{\ell(s), n_1(s), n_2(s), n_3(s)\}$ with $\ell(s) = \beta'(s)$ and $\langle \ell(s), \ell(s)\rangle = -\varepsilon$, satisfying the following Frenet equations:

$$\begin{aligned} \ell' &= \widetilde{k}_1 n_1, \\ n_1' &= \varepsilon \widetilde{k}_1 \ell + \widetilde{k}_2 n_2, \\ n_2' &= -\widetilde{k}_2 n_1 + \widetilde{k}_3 n_3, \\ n_3' &= -\widetilde{k}_3 n_2. \end{aligned} \tag{8.4}$$

Now we find how the curvature functions of Γ and Υ are linked.

Lemma 8.3.2 *Let Γ be a lightlike curve in the $\mathbb{L}^5$ and let Υ be the orthogonal projection of Γ onto a non-degenerate hyperplane Σ. Let us denote by t and s the pseudo-arc and arc parameters of Γ and Υ, respectively. Then s and t are linearly related if and only if the first curvature $\widetilde{k}_1$ of Υ is constant. Moreover, in this case, k_1 is constant if and only if $\widetilde{k}_2$ is constant.*

Proof In terms of the arc length parameter s Eq. (8.3) writes down as follows

$$\gamma(t(s)) = \beta(s) + \mu(s)\upsilon,$$

where $\beta(s) = \overline{\beta}(t(s))$ and $\mu(s) = \overline{\mu}(t(s))$. By taking derivatives with respect to s we have

$$L(t(s))t'(s) = \ell(s) + \mu'(s)\upsilon, \tag{8.5}$$

which implies

$$0 = -\varepsilon + \mu'(s)^2\varepsilon = \varepsilon(\mu'(s)^2 - 1),$$

and so $\mu(s)$ is a linear function. By derivating again we deduce

$$W_1(t(s))t'(s)^2 + L(t(s))t''(s) = \widetilde{k}_1 n_1,$$

that yields

$$t'(s)^4 = \widetilde{k}_1(s)^2.$$

This shows the first part of Lemma. The last claim of Lemma follows from the fact that $t''(s) = 0$ and $\widetilde{k}_1$ is constant. An easy and similar reasoning yields the conclusion. □

We are ready to get a relationship between lightlike Lancret helices in $\mathbb{L}^5$ with non-degenerate axis and non-degenerate curves in a hyperplane of $\mathbb{L}^5$.

Theorem 8.3.3 *Let Γ be a lightlike Cartan curve in $\mathbb{L}^5$, υ a constant unit vector and Σ the hyperplane orthogonal to υ in $\mathbb{L}^5$. Let Υ denotes the projection of Γ onto Σ. Then Γ is a Lancret helix with axis υ if and only if Υ is a curve in Σ with constant curvature and torsion.*

Proof From Lemma 8.3.2, to prove the first implication we only need to show that s and t are linearly related. From (8.5) we see that $\lambda t'(s) = \varepsilon\mu'(s) = \varepsilon$, where $\langle L, \upsilon\rangle = \lambda$ is constant, so we deduce that $t(s)$ is a linear function.

Conversely, let us assume that $\widetilde{k}_1$ and $\widetilde{k}_2$ are constant. From Lemma 8.3.2 we have that t and s are linearly related, and k_1 is constant. A straightforward computation leads to

$$k_1(t(s)) = -\frac{\widetilde{k}_2^2 - \varepsilon\widetilde{k}_1^2}{2\widetilde{k}_1},$$

$$k_2(t(s))^2 = \frac{\widetilde{k}_2^2}{\widetilde{k}_1^2}\widetilde{k}_3(s)^2,$$

$$k_3(t(s))^2 = \frac{1}{\widetilde{k}_1}\widetilde{k}_3(s)^2,$$

and hence $k_3 = rk_2$ with $r^2 = \widetilde{k}_1/\widetilde{k}_2^2$. □

As a consequence we have

Theorem 8.3.4 (Solving Natural Equations for Lancret Helices with Non-degenerate Axis) *Let Γ be a lightlike Cartan curve in $\mathbb{L}^5$. Then Γ is a Lancret helix with non-degenerate axis if and only if it is a lightlike geodesic of a Lorentzian cylinder constructed on a non-degenerate curve in $\mathbb{R}^4$ or $\mathbb{L}^4$ with constant curvature and torsion.*

8.3.3 The 5-Dimensional Case with Lightlike Axis

From now on, we will deal with lightlike Lancret helices with lightlike axis. The main difficulty now arises because of the orthogonal hyperplane to the axis is also lightlike and there is no unique way of projection.

Let v be the axis of the helix and let Σ be the orthogonal hyperplane, so that $v \in \Sigma$. From general theory of lightlike hypersurfaces (see [37]) we have the splitting

$$T_p\mathbb{L}^5 = T_p\Sigma \oplus tr(T_p\Sigma) = (\text{span}\{v\} \perp S(T_p\Sigma)) \oplus tr(T_p\Sigma), \quad \forall p \in \Sigma,$$

where $tr(T\Sigma) = \bigcup_{p\in\Sigma} tr(T_p\Sigma)$ is called a *screen transversal vector bundle* and $S(T\Sigma) = \bigcup_{p\in\Sigma} S(T_p\Sigma)$ is called a *screen distribution*. Then each choice of a screen distribution provides a projection map on Σ, so that the problem is finding a canonical screen distribution (or the most canonical screen distribution in some sense).

Let $\gamma : I \longrightarrow \mathbb{L}^5$ be a lightlike Lancret helix with lightlike axis v. Since $\langle L, v\rangle = \lambda$ is constant, then $\tilde{L} = \frac{1}{\lambda}L$ is a transversal section along γ satisfying $\langle \tilde{L}, v\rangle = 1$. Let $\overline{\beta}$ denotes the projection of γ with respect to $\tilde{L}$, which is given by

$$\overline{\beta}(t) = \gamma(t) - \langle\gamma(t), v\rangle\,\tilde{L}(t).$$

Since $\langle \gamma'(t), \upsilon \rangle = \lambda$, then $\langle \gamma(t), \upsilon \rangle = \lambda(t+\sigma)$ where σ is a constant, from which we have

$$\overline{\beta}(t) = \gamma(t) - (t+\sigma)L(t). \tag{8.6}$$

The Frenet equations (8.2) read as follows

$$\begin{aligned} T' &= \rho_1 V_1 \\ V_1' &= -\rho_1 T + \rho_2 V_2 \\ V_2' &= -\rho_2 V_1 + \rho_3 \upsilon \\ \upsilon' &= 0 \end{aligned} \tag{8.7}$$

where we have choosen $\mathcal{L} = \upsilon$.

Theorem 8.3.5 *Let γ be a lightlike Cartan curve in $\mathbb{L}^5$. If γ is a Lancret helix with lightlike axis υ and Σ denotes its lightlike orthogonal hyperplane, then the curvatures ρ_1, ρ_2 and ρ_3 of the projected spacelike curve $\overline{\beta}$ are given by:*

$$\rho_1(s) = \frac{\tilde{r}}{\sqrt{s}}, \quad \rho_3 = \tilde{r}_1 \rho_2, \tag{8.8}$$

for certain constants $\tilde{r}$ and $\tilde{r}_1$. Conversely, if Υ is a spacelike curve in a lightlike hyperplane of $\mathbb{R}^5_1$ whose curvatures satisfy (8.8), *then there exist a lightlike Lancret helix Γ in $\mathbb{L}^5$ whose projection onto Σ is just exactly Υ.*

Proof Let s denotes the arc length parameter of Υ, then (8.6) can be rewritten as

$$\beta(s) = \gamma(t(s)) - (t(s)+\sigma)L(t(s)),$$

where t stands for the pseudo-arc parameter of Γ. Differentiating and using the Frenet equations we get $T(s) = -(t(s)+\sigma)t'(s)W_1(t(s))$, from which we deduce that $t'(s) = 1/\sqrt{\sigma^2 + 2(s+\omega)}$, for a constant ω. Without loss of generality, let us assume that $\sigma = \omega = 0$. Differentiating again and using that k_1 is constant and $k_3 = r_1 k_2$, with r_1 constant, we deduce that

$$\rho_1(s) = \frac{\sqrt{k_1}}{\sqrt{s}}, \quad \rho_2(s) = \frac{k_2(s)}{2\sqrt{k_1}\sqrt{s}}, \quad \rho_3(s) = -\frac{r_1 k_2(s)}{\sqrt{2s}}.$$

Then we can take $\tilde{r} = \sqrt{k_1}$ and $\tilde{r}_1 = -\sqrt{2k_1}r_1$, and this concludes the proof.

Conversely, let $\beta : J \longrightarrow \Sigma$ be a spacelike curve in a lightlike hyperplane Σ. Put $\Sigma = \operatorname{span}\{\upsilon\}^{\perp}$, where υ is a lightlike vector. Let $\{T, V_1, V_2, \mathcal{L} = \upsilon\}$ be the Frenet frame satisfying (8.7), where the curvatures verify (8.8) (s denoting the arc length parameter). The Frenet frame $\{T, V_1, V_2, \mathcal{L} = \upsilon\}$ can be completed (in a unique way) to a basis of $T_{\beta(s)}\mathbb{L}^5$, for all $s \in J$, by adding a vector field $\mathcal{N}(s)$ along $\beta(s)$

satisfying

$$\langle \mathcal{L}, \mathcal{N} \rangle = -1, \quad \langle T, \mathcal{N} \rangle = \langle V_1, \mathcal{N} \rangle = \langle V_2, \mathcal{N} \rangle = \langle \mathcal{N}, \mathcal{N} \rangle = 0.$$

An easy computation shows that $\mathcal{N}'(s) = \rho_3(s)V_2(s) = \tilde{r}_1\rho_2(s)V_2(s)$. Then it is a straightforward computation to see that the curve

$$\overline{g}(s) = \beta(s) + \frac{\sqrt{s}}{\tilde{r}}\left(V_1(s) - \frac{\tilde{r}_1}{2}\upsilon - \frac{1}{\tilde{r}_1}\mathcal{N}(s)\right)$$

is a lightlike Lancret helix with axis υ. Now let t be the pseudo-arc parameter of $\overline{g}$ and put $\overline{g}(s) = \gamma(t(s))$. By taking derivatives we get

$$\overline{g}''(s) - \frac{1}{2s}T(s) + \frac{1}{4\tilde{r}s\sqrt{s}}\left(-V_1(s) + \frac{\tilde{r}_1}{2}\upsilon + \frac{1}{\tilde{r}_1}\mathcal{N}(s)\right),$$

from which we deduce $\langle \overline{g}''(s), \overline{g}''(s) \rangle = t'(s)^4 = 1/4s^2$, and so $t'(s) = 1/\sqrt{2s}$. From here, a long and messy computation yields the Frenet frame of $\overline{g}$:

$$\begin{aligned} L &= \frac{1}{\sqrt{2}\tilde{r}}\left(V_1 - \frac{\tilde{r}_1}{2}\upsilon - \frac{1}{\tilde{r}_1}\mathcal{N}\right), \\ W_1 &= -T, \\ N &= \frac{\tilde{r}}{\sqrt{2}}\left(V_1 - \frac{\tilde{r}_1}{2}\upsilon - \frac{1}{\tilde{r}_1}\mathcal{N}\right), \\ W_2 &= -V_2, \\ W_3 &= -\frac{\tilde{r}_1}{2} + \frac{1}{\tilde{r}_1}\mathcal{N}. \end{aligned}$$

These equations imply that $\overline{g}$ is a lightlike Lancret helix. Moreover, the curvatures are given by

$$k_1 = \tilde{r}^2, \quad k_2 = 2\tilde{r}\sqrt{s}\rho_2, \quad k_3 = \sqrt{2}\sqrt{s}\rho_2,$$

which concludes the proof. □

Now let γ be a lightlike Lancret helix with lightlike axis, whose curvatures satisfy $k_1(t) = r$ and $k_3(t) = r_1k_2(t)$, then the axis is given by

$$\upsilon = -\frac{1}{2}\left(rL + N + \frac{1}{r_1}W_3\right), \qquad r = \frac{1}{2r_1^2}.$$

Let us consider a timelike curve $\beta : J \longrightarrow \mathbb{L}^5$ with Frenet frame $\{\ell, n_1, n_2, n_3, n_4\}$ satisfying the following equations:

$$\begin{aligned} \ell' &= \widetilde{k}_1 n_1, \\ n_1' &= \widetilde{k}_1 \ell + \widetilde{k}_2 n_2, \\ n_2' &= -\widetilde{k}_2 n_1 + \widetilde{k}_3 n_3, \\ n_3' &= -\widetilde{k}_3 n_2 + \widetilde{k}_4 n_4, \\ n_4' &= -\widetilde{k}_4 n_3, \end{aligned} \tag{8.9}$$

with $\ell(s) = \beta'(s)$, s standing for the arc length parameter. Following [56], the curve Υ is a Lancret helix if there exist a non-zero constant vector υ such that the products $\langle \beta', \upsilon \rangle$ and $\langle n_2, \upsilon \rangle$ are constant. In this case, the Lancret theorem assures us that $\widetilde{k}_2 = \widetilde{r}_1 \widetilde{k}_1$ and $\widetilde{k}_4 = \widetilde{r}_3 \widetilde{k}_3$. Moreover, the axis is given by

$$\upsilon = \frac{1}{2} \left(\ell + \frac{1}{\widetilde{r}_1} n_2 + \frac{1}{\widetilde{r}_1 \widetilde{r}_3} n_4 \right),$$

where $\widetilde{r}_1 = \sqrt{1 + 1/\widetilde{r}_3^2}$ if the axis υ is a lightlike vector.

Let us consider the surface S locally parameterized by $X(s, \omega) = \beta(s) + \omega \upsilon$, then

$$\frac{\partial X}{\partial s}(s, \omega) = \ell(s) \qquad \text{and} \qquad \frac{\partial X}{\partial \omega} = \upsilon,$$

showing that S is a Lorentzian surface of $\mathbb{L}^5$. The lightlike geodesics in S can be parameterized by

$$\overline{g}(s) = \beta(s) - (s + \sigma)\upsilon, \tag{8.10}$$

where σ is constant. Let t be the pseudo-arc parameter of $\overline{g}$ (as a curve of $\mathbb{L}^5$), then $\overline{g}(s) = \gamma(t(s))$. A long and messy computation, from Eq. (8.10), yields the following relations among the Cartan curvatures of $\overline{g}$ and the Lancret helix β:

$$\begin{aligned} k_1(t(s)) &= \frac{1}{2} \frac{\widetilde{k}_1(s)}{\widetilde{r}_3^2} - \frac{7}{8} \frac{\widetilde{k}_1'(s)^2}{\widetilde{k}_1(s)^3} + \frac{1}{2} \frac{\widetilde{k}_1''(s)}{\widetilde{k}_1(s)^2}, \\ k_2(t(s)) &= \frac{\sqrt{1 + \widetilde{r}_3^2}}{\widetilde{r}_3} \widetilde{k}_3^2(s), \\ k_3(t(s)) &= \sqrt{\frac{1 + \widetilde{r}_3^2}{\widetilde{k}_1(s)}} \widetilde{k}_3(s). \end{aligned}$$

As a consequence, we obtain that $\overline{g}$ is a lightlike Lancret helix with lightlike axis if and only if $\widetilde{k}_1(s)$ is constant. In this case,

$$k_1 = r = \frac{1}{2}\frac{\widetilde{k}_1}{\widetilde{r}_3^2} \qquad \text{and} \qquad r_1 = \frac{\widetilde{r}_3}{\sqrt{\widetilde{k}_1}}.$$

From here and by using the theorem of existence and uniqueness of timelike curves $\mathbb{L}^5$, we can prove the following theorem.

Theorem 8.3.6 (Solving Natural Equations for Lancret Helices with Degenerate Axis) *Let Γ be a lightlike Cartan curve in $\mathbb{L}^5$. Then Γ is a Lancret helix with lightlike axis if and only if it is a geodesic in a Lorentzian ruled surface whose directrix is a timelike Lancret helix in $\mathbb{L}^5$ with lightlike axis and with constant first curvature, and whose rulings have the direction of the axis.*

Chapter 9
The Hasimoto Transformation

9.1 Rotational Magnetic Vector Fields

The rotational Killing vector fields have divergence zero, so they define magnetic vector fields in a Euclidean space. Thus, a natural problem is to know the dynamics of charges under the influence of those fields in $M = \mathbb{R}^3 - \{z - \text{axis}\}$. Since there are other divergence zero rotational vector fields which are not Killing, we have to set out the problem in a more accurate way. To do so, let us consider the natural action of the one-parameter group G generated by $V = \partial_\theta$ on M, i.e.,

$$G \times M \to M, \qquad (\phi_\theta, p) \mapsto \phi_\theta(p) = p \cdot R_\theta,$$

whose orbits are the integral curves of $V = \partial_\theta$ (they are circles centered on the z-axis lying in parallel planes to the plane $z = 0$). Furthermore, the orbit space can be identified with the half-plane $P = \{(x, 0, z) : x > 0\}$ and the quotient map $\pi : M \to P$ is given by

$$\pi(\rho \cos\theta, \rho \sin\theta, z) = (\rho, 0, z).$$

A domain $\Omega \subset M$ will be called *rotationally invariant*, or simply invariant, when it contains the orbit of any point $p \in \Omega$, that is, $\phi_\theta(p) \in \Omega$, for all $\theta \in \mathbb{R}$. It is clear that $\Omega = \pi^{-1}(U)$ is an invariant domain for any open set $U \subset P$ in the orbit space. Furthermore, any invariant domain is obtained in this way.

On the other hand, it is obvious that $V = \partial_\theta$ is a magnetic vector field on each invariant domain $\Omega = \pi^{-1}(U)$ and we wish to determine the functions $F \in C^\infty(\Omega)$ such that the field $F\,\partial_\theta$ remains as a magnetic vector field. To answer this question just remember that the following well known property of the divergence:

$$\operatorname{div}(F\,\partial_\theta) = F \operatorname{div}(\partial_\theta) + \langle \nabla F, \partial_\theta \rangle = \langle \nabla F, \partial_\theta \rangle = \partial_\theta(F),$$

A. Ferrandez Izquierdo, *Helices*, RSME Springer Series 16,
https://doi.org/10.1007/978-3-032-09663-0_9

which shows that the divergence of $F\partial_\theta$ vanishes if and only if $\partial_\theta(F) = F_\theta = 0$, and that is equivalent to $F = f \cdot \pi$, where $f \in C^\infty(U)$. From now on, these fields will be called *rotational magnetic vector fields*.

We aim to study the dynamics of charges in invariant domains under the influence of rotational magnetic vector field. In other words, we wish to find the solutions of the Lorentz force equation for rotational magnetic vector fields. We focus on the solutions of the equation

$$\gamma''(s) = (f \cdot \pi)\, \partial_\theta \times \gamma'(s), \quad \text{in} \quad \Omega = \pi^{-1}(U), \quad \text{with} \quad f \in C^\infty(U). \tag{9.1}$$

The first approach to the study of that differential equation comes from the following geometric characterization of their solutions:

A curve in $\mathbb{R}^3$ is a magnetic curve of a rotational magnetic vector field if and only if it is a geodesic of a revolution surface having the same axis as the magnetic vector field.

In fact, let $\beta(s) = (\rho(s), 0, z(s))$ be a plane curve which is rotated around the z-axis to obtain the revolution surface S_β. Let us consider an arc length parameterized geodesic $\gamma(s) = (\rho(s)\cos\theta(s), \rho(s)\sin\theta(s), z(s))$ of that surface. It is clear that

$$\gamma(s) = \phi_{\theta(s)}(\beta(s)) = \beta(s) \cdot R_{\theta(s)},$$

so that $\gamma'(s) = \beta'(s).R_{\theta(s)} + \theta'(s)\, \partial_\theta(\gamma(s))$. In particular,

$$1 = |\gamma'(s)|^2 = |\beta'(s)|^2 + \rho^2(\theta')^2, \tag{9.2}$$

which is combined with the Clairaut condition

$$\rho^2\theta' = r, \tag{9.3}$$

to find that

$$|\beta'(s)|^2 + \frac{r_1^2}{\rho^2} = 1. \tag{9.4}$$

The tangent plane of S_β, along $\gamma(s)$ is generated by the vector fields $\phi_{\theta(s)}(\beta'(s)) = \beta'(s) \cdot R_{\theta(s)}$ and $\partial_\theta(\gamma(s))$. Therefore,

$$\partial_\theta(\gamma(s)) \times \gamma'(s) = \partial_\theta(\gamma(s)) \times \phi_{\theta(s)}(\beta'(s)),$$

is normal to S_β along $\gamma(s)$. However, if $\gamma(s)$ is a geodesic of S_β, then $\gamma''(s)$ is also normal to S_β along $\gamma(s)$, which implies the existence of a certain function $f(s)$ satisfying

$$\gamma''(s) = f(s)\, \partial_\theta(\gamma(s)) \times \gamma'(s).$$

Naturally, the function f is defined along the profile curve $\beta(s)$, and so $f \circ \pi$ is a function on the revolution surface S_β. Now f can be extended to a neighborhood (for instance, a tubular one) U of the profile curve $\beta(s)$ in the half-plane P. Such extension, say $\overline{f}$, defines the potential $\overline{f} \circ \pi$ with $\partial_\theta(\overline{f}) = 0$, in the invariant domain $\Omega = \pi^{-1}(U)$, which means that $\gamma(s)$ is a solution of the corresponding Lorentz force equation (9.1).

As for the converse, let $\gamma(s) = (\rho(s)\cos\theta(s), \rho(s)\sin\theta(s), z(s))$ a solution of (9.1), in a certain invariant domain, which can be assumed to be arc length parameterized. Then

$$\gamma(s) = \phi_{\theta(s)}(\beta(s)), \qquad \beta(s) = (\rho(s), 0, z(s)),$$

and we can consider the revolution surface with profile curve $\beta(s)$

$$S_\beta \, : \, X(s,\theta) = \phi_\theta(\beta(s)).$$

It is clear that $\gamma(s) = X(s, \theta(s))$, so that the given curve is contained in S_β. Moreover, as $\gamma(s)$ is a solution of (9.1), we have that

$$\gamma''(s) = (f \cdot \pi)\, \partial_\theta \times (\phi_{\theta(s)}(\beta'(s)) + \theta'\, \partial_\theta) = (f \cdot \pi)\, \partial_\theta \times \phi_{\theta(s)}(\beta'(s)),$$

which shows that it is normal to S_β along $\gamma(s)$, and therefore it is a geodesic of S_β.

An important consequence is the following:

The dynamics of charges under the action of a rotational magnetic vector field is equivalent to the gravity action in revolution surfaces having the same axis as the field. In other words, a charge submitted to the action of a rotational magnetic vector field evolves as a free fall particle in a revolution surface having the same axis as the vector field.

9.2 Some Examples and Applications

Just as we did with translational magnetic vector fields, whose magnetic curves are Lancret helices, but now taking rotational magnetic vector fields, the Lorentz force equation (9.1) allows us to compute its potential along any magnetic curve as follows:

$$\langle \gamma'', \gamma'' \rangle = \kappa^2(s) = f^2(s)\, \rho^2(s)\, \sin^2\varphi(s),$$

where $\varphi(s)$ denotes the angle that the magnetic curve makes with the rotational vector field, i.e., with parallels of the revolution surface. Furthermore, $\kappa(s)$ is the curvature in $\mathbb{R}^3$ of the magnetic curve, which obviously agree with its normal curvature in the revolution surface where is lying as a geodesic. Therefore, on any magnetic curve the potential f is given by

$$f(s) = \frac{\kappa(s)}{\rho(s)\,\sin\varphi(s)}. \tag{9.5}$$

As a consequence we obtain that *any plane curve is a magnetic curve of a rotational magnetic vector field.* In fact, let $\beta(s)$ be a curve in a plane Π in $\mathbb{R}^3$ and let L be a straight line in Π which not meet the curve. To fix ideas and without loss of generality, take L as the z-axis and let Π be the plane of equation $y = 0$, so that β can be considered in the half-plane $P = \{(x, 0, z) : x > 0\}$. By rotating the curve around L we obtain a revolution surface S_β whose meridians are geodesics. Then we state the problem of looking for potentials f such that the meridians of S_β are magnetic curves of the magnetic vector field $V = f\,\partial_\theta$. The only condition that f should satisfy is given by the formula (9.5), i.e., along the curve $\beta(s)$ the potential must be of the form

$$f(s) = \frac{\kappa(s)}{\rho(s)}, \tag{9.6}$$

where κ is the curvature of β, which geometrically means that the potential is the product of the curvatures, in $\mathbb{R}^3$, of the meridian and parallel, respectively, at any point. Now the function $f(s)$, defined along $\beta(s)$, must be extended to a neighborhood of the curve in the half-plane P to obtain the potential of a rotational magnetic vector field which admits the meridians of S_β as magnetic curves. Before to go on, we point out the following:

Remark If a magnetic curve $\gamma(s)$ of a rotational magnetic vector field, $V = f\,\partial_\theta$, is orthogonal at a point to such a field, then the magnetic curve is a meridian, and therefore a plane curve in a revolution surface. To understand that, we know that $\gamma(s)$ is a geodesic of a certain revolution surface S_β. Then if $p = \gamma(0)$ is the point where they are orthogonal, and $v = \gamma'(0)$, the meridian $\delta(s)$ throughtout p and the given magnetic curve are two geodesics of S_β with the same initial conditions (p, v), and then they agree.

We will see some applications.

(1) By removing the z-axis from $\mathbb{R}^3$ we have a space that can be foliated by concentric spheres centered at the origen. More precisely, let us start with an open semicircle

$$\beta_a(s) = (\rho(s), 0, z(s)) = \left(a\,\cos\frac{s}{a}, 0, a\,\sin\frac{s}{a}\right), \qquad s \in \left(-\frac{\pi a}{2}, \frac{\pi a}{2}\right).$$

Now rotate it around the z-axis to obtain a sphere with two antipodal points removed, $S_a = \mathbb{S}^2(a) - \{p_1, p_2\}$, centered at the origin and radius a. Then we have an invariant domain

$$\Omega = \mathbb{R}^3 - \{z - \text{axis}\} = \bigcup_{a>0} S_a.$$

We are going to find a rotational magnetic vector field $V = f\,\partial_\theta$ in Ω such that its magnetic curves, which are orthogonal to that field, are the meridians of the spheres of the above foliation. To get it, use (9.6), which in this case is written as

$$f(s) = \frac{1}{a\,\rho(s)}.$$

Now we consider that $a = \sqrt{\rho^2 + z^2}$ to obtain the potential

$$f : \Omega \to \mathbb{R}, \qquad f(\rho, \theta, z) = \frac{1}{\rho\sqrt{\rho^2 + z^2}},$$

that solves the problem. Summing up, *the orthogonal magnetic flow generated by the rotational vector field*

$$V(\rho, \theta, z) = \frac{1}{\rho\sqrt{\rho^2 + z^2}}\,\partial_\theta$$

is made up of meridians of one of the above family of concentric spheres with respect to the origin.

(2) As a second application, we wish to find rotational vector fields in suitable invariant domains such that the meridians of a foliation by revolution tori, in that domain, are magnetic curves. Let us start from a revolution torus with radii $R > a$, which is obtained by rotating a circle centered at $(R, 0, 0)$ with radius a around the z-axis, that is, the profile curve is

$$\beta_a(s) = (\rho(s), 0, z(s)) = \left(R + a\cos\frac{s}{a}, 0, a\sin\frac{s}{a}\right),$$

and the revolution torus is

$$S_a : \quad X_a(s, v) = \left(\left(R + a\cos\frac{s}{a}\right)\cos v, \left(R + a\cos\frac{s}{a}\right)\sin v, a\sin\frac{s}{a}\right).$$

From (9.6), the meridians of this torus are magnetic curves of all rotational vector fields with potentials such that restricted to β_a have the form

$$f(s) = \frac{1}{a\,\rho(s)} = \frac{1}{a\left(R + a\cos\frac{s}{a}\right)}.$$

Let us consider the open disc $U = \{(x, 0, z) : 0 < (x - R)^2 + z^2 < R\}$, in the half-plane P, with radius R and centered at $(R, 0, 0)$, which we rotate around the z-axis to get an invariant domain $\Omega = \pi^{-1}(U) \subset \mathbb{R}^3$. Then Ω is foliated by revolution tori with radii R and $a < R$, that is,

$$\Omega = \bigcup_{0<a<R} S_a.$$

Now we define the potential

$$f : \Omega \subset \mathbb{R}^3 \to \mathbb{R}, \qquad f(\rho, \theta, z) = \frac{1}{\rho\sqrt{(\rho - R)^2 + z^2}},$$

which gives the rotational vector field

$$V(\rho, \theta, z) = \frac{1}{\rho\sqrt{(\rho - R)^2 + z^2}}\, \partial_\theta,$$

whose orthogonal magnetic flow is made up of the meridians of the above foliation of Ω.

(3) The half-plane P can be foliated by a family of circles tangent at the origin (which was just removed). More precisely, take the punctured circles

$$\beta_a(s) = (\rho(s), 0, z(s)) = \left(a + a\cos\frac{s}{a}, 0, a\sin\frac{s}{a}\right), v \qquad a > 0,$$

and rotate them around the z-axis to obtain revolution tori tangent at the origin, previously removed. We are going to find rotational magnetic vector fields whose orthogonal magnetic flow agree with the family of medians of that foliation in $\mathbb{R}^3 - \{z - \text{axis}\}$. Using again (9.6), we have

$$a + a\cos\frac{s}{a} = \rho, \qquad a\sin\frac{s}{a} = z,$$

so that

$$a^2 = (\rho - a)^2 + z^2,$$

and

$$a = \frac{\rho^2 + z^2}{2\rho}.$$

The solution is the following rotational magnetic vector field

$$V(\rho, \theta, z) = \frac{2}{\rho^2 + z^2}\, \partial_\theta .$$

(4) Let us do an ambitious proposal. Given a revolution surface S_β, find the rotational magnetic vector field $V = f\, \partial_\theta$ such that the geodesics of S_β are contained in the magnetic flow. We will see that this problem has no solution, in other words, *in general, there is no rotational magnetic vector field admitting as magnetic curves the geodesics of the given surface*. First, observe that the potential of a rotational magnetic vector field, along a magnetic curve $\gamma(s)$ is given by (9.5). Then, take, for instance, the foliation by spheres in $\Omega = \mathbb{R}^3 - \{z - \text{axis}\}$ given in Application 1. The potential of magnetic curves (maximal circles) satisfies that

$$f(s) = \frac{1}{a\, \rho(s)\, \sin\varphi(s)},$$

so that we have to control the angle $\varphi(s)$ that makes with the flow of parallels (tangent to rotational vector field). To do that, we use the Clairaut formula (9.3), which can be written as

$$\rho(s)\, \cos\varphi(s) = r. \tag{9.7}$$

From here it is easy to obtain that

$$\sin\varphi(s) = \frac{\sqrt{\rho(s)^2 - r^2}}{\rho(s)},$$

and so the potential is

$$f(s) = \frac{1}{a\, \sqrt{\rho(s)^2 - r^2}}, \tag{9.8}$$

where r is the Clairaut constant of the maximal circle $\gamma(s)$ in the corresponding sphere. As a consequence, we obtain a potential, and then a rotational magnetic vector field whose flow contains the geodesics having the same Clairaut constant. More precisely, the rotational magnetic vector field above defined in Ω, given by

$$V(\rho, \theta, z) = \frac{1}{\sqrt{(\rho^2 - r^2)(\rho^2 + z^2)}}\, \partial_\theta,$$

contains in its flow all geodesics with Clairaut constant r in the foliation by concentric spheres.

(5) The geodesics of a catenoid were explicitly obtained in [90]. For a positive constant $a > 0$, let us consider the catenary $\beta(t) = (a\cosh t, 0, at)$ in the half-plane P which will be rotated around the z-axis to get the catenoid S_β parameterized by

$$X(t, \theta) = (a\cosh t\ \cos\theta, a\cosh t\ \sin\theta, at).$$

The normal curvature, κ_n, of β in S_β, which agrees with its curvature in the half-plane P, is given by

$$\kappa_n = -\frac{1}{a\cosh^2 t}.$$

Now take any geodesic, γ, of the catenoid, and set r its Clairaut constant. Then we can compute its curvature in $\mathbb{R}^3$, which will agree with its normal curvature in the catenoid. We will use the classical Euler formula and the fact that the catenoid is a minimal surface in $\mathbb{R}^3$ to deduce that

$$\kappa = \kappa_n(\cos^2\varphi - \sin^2\varphi) = \frac{-2r^2 + a^2\cosh^2 t}{a^3\cosh^4 t},$$

where φ stands for the angle that γ makes with ∂_θ. Now, use again (9.5) to check that the given geodesic is a magnetic curve of the following rotational magnetic vector field

$$V = \frac{a(\rho^2 - 2r^2)}{\rho^4\sqrt{\rho^2 - r^2}}\,\partial_\theta.$$

9.3 The Lorentz Force Equation of Rotational Magnetic Vector Fields

As we have just seen, the study of translational magnetic vector fields $f(x, y)\,\partial_z$ in $\mathbb{R}^3$ has been completely understood. The corresponding magnetic curves are geodesics in right cylinders built on plane curves, in other words, they are Lancret helices, and therefore the complete integration of the Lorentz force equations can be easily obtained (see [7] and [9]).

On the contrary, the study of rotational magnetic vector fields is more complicated and subtle, because, as we saw, their magnetic trajectories appear as geodesics of revolution surfaces and the theory of this kind of curves is not yet well understood.

In this section we will express the Lorentz force equation of a rotational magnetic vector field in cylindrical coordinates. To do that and without loss of generality, we will take the z-axis as the axis of the field and use the cylindrical coordinates $(x, y, z) = (\rho\cos\theta, \rho\sin\theta, z)$ so that we obtain the rotational magnetic vector field $f(\rho, z)\,\partial_\theta$. A straightforward computation allows us to see that the corresponding Lorentz force equation (9.1) writes down as the following system of differential equations

$$\begin{aligned}(\rho'' - \rho(\theta')^2 - f\rho z')\cos\theta - (2\rho'\theta' + \rho\theta'')\sin\theta &= 0,\\ (2\rho'\theta' + \rho\theta'')\cos\theta + (\rho'' - \rho(\theta')^2 - f\rho z')\sin\theta &= 0,\\ z'' &= -f\rho\rho',\end{aligned}$$

which can be written as

$$\begin{aligned}\rho'' - \rho(\theta')^2 - f\rho z' &= 0,\\ 2\rho'\theta' + \rho\theta'' &= 0,\\ z'' &= -f\rho\rho'.\end{aligned}$$

Nevertheless, as we know that its solutions are geodesics in revolution surfaces, the second equation can be replaced by Clairaut equation (9.10). Thus, the magnetic curves of the rotational magnetic vector field $f(\rho, z)\,\partial_\theta$ are the solutions of the following system of differential equations

$$\rho'' - \frac{r_1^2}{\rho^3} - f\rho z' = 0, \tag{9.9}$$

$$\rho^2(s)\,\theta'(s) = r_1, \tag{9.10}$$

$$z'' = -f\rho\rho'. \tag{9.11}$$

We wish to point out that the solutions, $\gamma(s) = (\rho(s)\cos\theta(s), \rho(s)\sin\theta(s), z(s))$, are curves with constant speed, i.e., they are automatically arc length parameterized. In fact,

$$\rho'' - \frac{r_1^2}{\rho^3} + \frac{z'z''}{\rho'} = 0,$$

which can be written as

$$\left[(\rho')^2 + (z')^2 + \frac{r_1^2}{\rho^2}\right]' = (|\gamma'(s)|^2)' = 0.$$

It is clear that the full integration of this system, for an arbitrary potential $f(\rho, z)$ is not possible. However, some information about the solutions can be obtained for appropriate potentials (see [17]). In the next section we will consider the case of a constant potential, which is equivalent to take a Killing magnetic vector field.

9.4 Elastic and Rotational Magnetic Curves

A constant potential provides rotational magnetic vector fields that are Killing. Without loss of generality, set $f = 1$, so that we are talking about dynamics of charges associated to the field ∂_θ. In this case, the Eqs. (9.9) and (9.11) are written as

$$\rho'' - \frac{r_1^2}{\rho^3} - \rho z' = 0, \tag{9.12}$$

$$z'' = -\rho\rho', \tag{9.13}$$

r_1 being a constant, which we will call the Clairaut constant (9.10). Then, the magnetic trajectories are the curves $\gamma(s) = (\rho(s)\cos\theta(s), \rho(s)\sin\theta(s), z(s)$ which are solutions of the following system of differential equations

$$\rho'' + \frac{1}{2}\rho^3 - A_o\rho - \frac{r_1^2}{\rho^3} = 0, \tag{9.14}$$

$$\rho^2\theta' = r_1 \text{ (constant)}, \tag{9.15}$$

$$z' = -\frac{1}{2}\rho^2 + A_o, \tag{9.16}$$

A_o standing for an arbitrary constant obtained when integrating (9.13). Those equations can be fully integrated by using elliptic functions (see [8, 14] and [36]). However, we will offer a more subtle procedure, used in [17], to see that the solutions of the above system were already essentially known. To do that, first observe that these solutions are closely related to those of the Daniel Bernoulli classical variational problem of elastic curves. According to that model, an elastic is a curve in $\mathbb{R}^3$ which is an extremal of the total squared curvature acting on the space of curves with the same length and satisfying some first order boundary conditions. In other words, the elastic curves are the critical points of the following energy functional

$$\mathcal{F}_\lambda(\alpha) = \int_\alpha (\kappa^2 + \lambda)\, ds \tag{9.17}$$

acting on the space of curves satisfying appropriate forst order boundary conditions (the so called clamped curves), where λ plays the role of a Lagrange multiplier. The Euler-Lagrange equations associated to that elastic energy can be computed by standard techniques involving integrations by parts (more details can be seen in [69] and [70]). Therefore, the elastic curves in $\mathbb{R}^3$ are those ones whose curvature functions κ and τ are solutions of the following Euler-Lagrange equations

$$\kappa'' + \frac{1}{2}\kappa^3 - \kappa\tau^2 - \frac{\lambda}{2}\kappa = 0, \tag{9.18}$$

$$\kappa^2\tau = c \text{ (constant)}. \tag{9.19}$$

These equations can be easily manipulate to see that the above system can be rewritten as

$$\kappa'' + \frac{1}{2}\kappa^3 - \frac{\lambda}{2}\kappa - \frac{c^2}{\kappa^3} = 0, \tag{9.20}$$

$$\kappa^2\tau = c \text{ (constant)}. \tag{9.21}$$

We can see that that system, whose solutions give the curvature and torsion functions of the elastic curves in $\mathbb{R}^3$, is essentially the same as that composed by the Eqs. (9.14) and (9.15). Then we can draw the following conclusions:

1. Let $\alpha(s)$ be an arc length elastic curve in $\mathbb{R}^3$. Then there exists two parameters (λ, c) such that the curvature $\kappa(s) > 0$ and torsion $\tau(s)$ of $\alpha(s)$ are solutions of (9.20) and (9.21). Then we can build a curve $\gamma(s) = (\rho(s)\cos\theta(s), \rho(s)\sin\theta(s), z(s))$ in $\mathbb{R}^3 - \{z - \text{axis}\}$ as follows

$$\rho(s) = \kappa(s), \qquad \theta(s) = \int_0^s \tau(u)\,du, \qquad z(s) = \frac{1}{2}\int_0^s (\lambda - \kappa^2(u))\,du. \tag{9.22}$$

   We have found an arc length parameterized curve which is a solution of the system of differential equations (9.14), (9.15) and (9.16), with $r_1 = c$ and $A_o = \lambda/2$ (remember that all solutions of that system are arc length parameterized curves), and therefore a magnetic trajectory of the rotational magnetic vector field ∂_θ.
2. The above construction can be done in two steps. First, we build the projection of the curve on the plane $z = 0$, i.e.,

$$\left(\rho(s) = \kappa(s), \theta(s) = \int_0^s \tau(u)\,du, 0\right).$$

   Thus we get a solution of (9.14) and (9.15). Second, we define the height on that plane

$$z(s) = \frac{1}{2}\int_0^s (\lambda - \kappa^2(u))\,du,$$

   to find a solution of (9.16).

3. The solution (9.22), as any magnetic trajectory of a rotational magnetic vector field, can be viewed as a geodesic of a revolution surface. In this case it is a geodesic with Clairaut constant c in the revolution surface obtained by rotating, around the z-axis, the profile curve, in the plane $y = 0$, given by

$$\beta(s) = \left(\rho(s) = \kappa(s), 0, z(s) = \frac{1}{2} \int_0^s (\lambda - \kappa^2(u))\, du \right).$$

   This is a nice geometric interpretation of the constant c, which appears when obtaining the first integral (9.21) of the elastic curves. In fact, these constants are precisely the Clairaut ones when one sees the corresponding magnetic trajectories as geodesics in the corresponding revolution surfaces.
4. All magnetic trajectories of the field ∂_θ are obtained from elastic curves according to the above algorithm. In any case, the correspondence between elastic curves and rotational magnetic trajectories needs a little more detail, which we will provide in the coming sections.

9.5 The Hasimoto Transformation

The *Hasimoto transformation* $\mathcal{H}$ is, in principle, defined on the space C of connected curves in $\mathbb{R}^3$ without inflection points (see [53] and [54]). It associates to a curve $\alpha \in \mathrm{C}$ a function with complex values with no zeros, the so called *complex curvature* or *complex wave function* of the given curve. More precisely, let $\mathcal{H} : \mathrm{C} \to \mathrm{F}^o = \mathrm{F}(\mathbb{R}, \mathbb{C})$ be the map defined by

$$\mathcal{H}(\alpha(s)) = \phi(s), \quad \text{where} \quad \phi(s) = \kappa(s) \exp\left(i \int_0^s \tau(u)\, du \right), \tag{9.23}$$

where $\kappa(s) > 0$ and $\tau(s)$ stand, respectively, for the curvature and torsion functions of α.

It is clear that this map is actually defined on the space $\overline{\mathrm{C}}$ of congruence classes of curves of C, because it is invariant under rigid motions of $\mathbb{R}^3$.

We are going to find a one-to-one correspondence from de above Hasimoto transformation. To do that, we start by studying its presumed injectivity. Let α_1 and α_2 be two curves in C with the same image by $\mathcal{H}$. Then they have the same complex curvature

$$\phi_1(s) = \kappa_1(s) \exp\left(i \int_0^s \tau_1(u)\, du \right) = \phi_2(s) = \kappa_2(s) \exp\left(i \int_0^s \tau_2(u)\, du \right),$$

so that we find that

$$|\phi_1(s)| = |\phi_1(s)| \quad \Longrightarrow \quad \kappa_1(s) = \kappa_2(s).$$

Furthermore,

$$\exp\left(i\int_0^s (\tau_1(u) - \tau_2(u))\,du\right) = 1,$$

which means that

$$\tau_1(s) = \tau_2(s).$$

As both curves have the same curvature and torsion functions, they are congruent, and so they define the same congruence class. In other words, the Hasimoto transformation is injective provided it is defined on $\overline{\mathrm{C}}$.

As for surjectivity, we do the following test. Let us consider the image by $\mathcal{H}$ of a plane curve. Since $\tau = 0$, we have that its complex curvature is a real function

$$\phi(s) = \kappa(s).$$

In particular, the images by $\mathcal{H}$ of circles are fixed points lying on the real axis. This implies that when considering in F^o a constant in a non-real complex number, then that map can not be the image by $\mathcal{H}$ of any curve. In fact, take the map

$$\psi \in \mathrm{F}^o, \qquad \psi(s) = \kappa_o\, e^{i\theta_o},$$

where κ_o is a positive real number and θ_o a non-zero real number. By assuming that this map belongs to the image by $\mathcal{H}$, then it should be the complex curvature of a certain curve $\alpha(s)$, that is,

$$\phi(s) = \kappa(s)\exp\left(i\int_0^s \tau(u)\,du\right) = \psi(s) = \kappa_o\, e^{i\theta_o}.$$

This would imply that $\kappa(s) = \kappa_o$ and

$$\int_0^s \tau(u)\,du = \theta_o.$$

As θ_o is constant, then it should be zero, which is a contradiction with the choice of θ_o. As a consequence, the Hasimoto transformation $\mathcal{H} : \mathrm{C} \to \mathrm{F}^o$ is injective, but not surjective. However, it yields to a one-to-one transformation by defining an equivalence relation in F^o. To do that, we will consider the following natural action of $\mathbb{S}^1$ on F^o

$$\mathbb{S}^1 \times \mathrm{F}^o \to \mathrm{F}^o, \qquad e^{it} \cdot \phi(s) = \rho(s) \cdot e^{i(\theta(s)+t)}$$

where $\phi(s) = \rho(s) \cdot e^{i\theta(s)}$. Then we can see that the congruence class of a curve can be recovered from the equivalence class of its complex curvature $[\phi(s)]$ in $\mathrm{F}^o/\mathbb{S}^1$.

In fact, curvature κ and torsion τ can be expressed in terms of $[\phi(s)]$ as follows:

$$\kappa(s) = |\phi(s)|, \qquad \tau(s) = \frac{d}{ds}\arg[\phi(s)].$$

Then the Hasimoto transformation can be bijectively defined by

$$\mathcal{H} : \overline{\mathrm{C}} \to \mathrm{F}^o/\mathbb{S}^1, \qquad (\mathcal{H}(\alpha))(s) = [\phi(s)]. \tag{9.24}$$

On the other hand, F^o can be seen as the space of curves in the plane, with the origin removed. Then $\mathrm{F}^o/\mathbb{S}^1$ is the space of such curves, up to rotations around the origin. From now on, we will work in the quotient spaces $\overline{\mathrm{C}}$ and $\mathrm{F}^o/\mathbb{S}^1$, and the Hasimoto transformation will provide a one-to-one correspondence between connected curves in $\mathbb{R}^3$, with no inflection points, and plane curves which do not pass trough the origin ([53] and [54]).

We know that given a curve $\alpha \in \mathrm{C}$ we have its Frenet frame $\{T, N, B\}(s)$ whose variation is described by its curvature functions, $\kappa(s) > 0$ and $\tau(s)$, according to the Frenet equations

$$\frac{d}{ds}\begin{pmatrix} T(s) \\ N(s) \\ B(s) \end{pmatrix} = \begin{pmatrix} 0 & \kappa(s) & 0 \\ -\kappa(s) & 0 & \tau(s) \\ 0 & -\tau(s) & 0 \end{pmatrix}\begin{pmatrix} T(s) \\ N(s) \\ B(s) \end{pmatrix}.$$

We can obtain equivalent equations governed by the complex curvature. To do that, in the normal plane along the curve, we consider complex coordinates by defining the vector fields

$$\eta(s) = e^{i\int_0^s \tau(u)\,du}(N(s) + iB(s)),$$

$$\overline{\eta}(s) = e^{-i\int_0^s \tau(u)\,du}(N(s) - iB(s)).$$

Now we are going to study the variation of the Hasimoto frame $\{T(s), \eta(s), \overline{\eta}(s)\}$ along $\alpha(s)$.

First, from the first Frenet equation $T'(s) = \kappa(s)\,N(s)$ this field is in the normal plane along the curve, so that to write it in terms of $\{\eta, \overline{\eta}\}$ we realize that

$$\frac{1}{2}\overline{\phi}\,\eta + \frac{1}{2}\phi\,\overline{\eta} = \kappa\,N.$$

Furthermore, a straightforward computation from the Frenet equations yields

$$(N + iB)' + i\tau(N + iB) = -\kappa T,$$

so that

$$\begin{aligned}\eta' &= i\tau\, e^{i\int_0^s \tau(u)\,du}(N+iB) + e^{i\int_0^s \tau(u)\,du}(N+iB)' = -\kappa\, e^{i\int_0^s \tau(u)\,du} T \\ &= -\phi\, T.\end{aligned}$$

Then $\overline{\eta}' = -\overline{\phi}\, T$ and the variation of the Hasimoto frame along the curve is governed by the complex curvature $\phi(s)$ according to the following equations

$$\frac{d}{ds}\begin{pmatrix} T(s) \\ \eta(s) \\ \overline{\eta}(s)\end{pmatrix} = \begin{pmatrix} 0 & \frac{1}{2}\overline{\phi}(s) & \frac{1}{2}\phi(s) \\ -\phi(s) & 0 & 0 \\ -\overline{\phi}(s) & 0 & 0 \end{pmatrix}\begin{pmatrix} T(s) \\ \eta(s) \\ \overline{\eta}(s)\end{pmatrix}. \tag{9.25}$$

9.6 Hasimoto Computations: From the Filament to the Schrödinger Equations

We can often find curves α in $\mathbb{R}^3$ evolving over time tracing out surfaces of the following form

$$\alpha(s,t) = \varphi_t(\alpha(s)), \qquad \alpha(s,0) = \alpha(s).$$

In this context, we can find two kind of problems:

(i) We know the evolution mechanism, i.e., the transformations $\{\varphi_t \,:\, t \in \mathbb{R}\}$, and then we look for curves whose evolution yields a surface satisfying an appropriate property. For instance, we can think that the evolution is made trough rotations around an axis so that we ask for plane curves such that the corresponding surface is minimal.

(ii) We can also think that the surface $\alpha(s,t)$, obtained after a curve has evolved, satisfies a certain differential equation, usually known as the evolution equation, and then we try to find solutions of that equation. In this case, if we find a solution $\alpha(s,t)$, then the curve $\alpha(s)$, as generating curve, is often called the initial condition of the equation.

An interesting example of the latter situation can be described as follows. First, a *Hasimoto surface* $\alpha(s,t)$ is the surface traced out by a curve $\alpha(s)$ of C as it evolves over time in $\mathbb{R}^3$, according to the following evolution equation

$$\frac{\partial \alpha}{\partial t} = \frac{\partial \alpha}{\partial s} \times \frac{\partial^2 \alpha}{\partial s^2}. \tag{9.26}$$

This equation plays an important role in fluid dynamics, because it describes the motion of an isolated vortex filament in an incompressible fluid. This is why it is usually called the filament equation or localized induction equation (LIE). It was first formulated by L. Da Rios en 1906 (see [91]) and later by R. Betchov [23], so that it is also known as the *Betchov-Da Rios equation*. Then Hasimoto invented a transformation to show the following interesting result:

A curve of C **evolves under** LIE **if and only if its complex phase function is a solution of the following differential equation, known as the non-linear cubic Schrödinger equation**

$$i\phi_t + \phi_{ss} + \frac{1}{2}(|\phi|^2 + C)\phi = 0. \tag{9.27}$$

In other words, the study of Hasimoto surfaces is equivalent to that of Schrödinger equation.

To prove that, we start by considering the Hasimoto frame along the curves of the evolution, that is,

$$\eta(s,t) = e^{i\int_0^s \tau(u,t)\,du}(N(s,t) + iB(s,t)),$$

$$\overline{\eta}(s,t) = e^{-i\int_0^s \tau(u,t)\,du}(N(s,t) - iB(s,t)).$$

Then, from (9.25) we deduce the following longitudinal variation of that frame

$$\frac{d}{ds}\begin{pmatrix} T(s,t) \\ \eta(s,t) \\ \overline{\eta}(s,t) \end{pmatrix} = \begin{pmatrix} 0 & \frac{1}{2}\overline{\phi}(s,t) & \frac{1}{2}\phi(s,t) \\ -\phi(s,t) & 0 & 0 \\ -\overline{\phi}(s,t) & 0 & 0 \end{pmatrix}\begin{pmatrix} T(s,t) \\ \eta(s,t) \\ \overline{\eta}(s,t) \end{pmatrix}. \tag{9.28}$$

Let us compute now the transversal variation of the frame. First we have that

$$\frac{\partial \eta}{\partial t} = i\frac{\partial}{\partial t}\left(\int_0^s \tau(u,t)\,du\right)\eta + \left(e^{i\int_0^s \tau(u,t)\,du}\right)\frac{\partial}{\partial t}(N + iB) = i\,R\,\eta + \varphi\,T,$$

R being a real function, whereas φ is a certain complex function. Then it is clear that

$$\frac{\partial \overline{\eta}}{\partial t} = \overline{\varphi}\,T - i\,R\,\overline{\eta}.$$

We can also see that

$$\frac{\partial T}{\partial t} = -\frac{1}{2}\overline{\varphi}\,\eta - \frac{1}{2}\varphi\,\overline{\eta}.$$

Summarizing, we write in a matricial form as

$$\frac{d}{dt}\begin{pmatrix} T(s,t) \\ \eta(s,t) \\ \overline{\eta}(s,t) \end{pmatrix} = \begin{pmatrix} 0 & -\frac{1}{2}\overline{\varphi}(s,t) & -\frac{1}{2}\varphi(s,t) \\ \varphi(s,t) & i\,R & 0 \\ \overline{\varphi}(s,t) & 0 & -i\,R \end{pmatrix}\begin{pmatrix} T(s,t) \\ \eta(s,t) \\ \overline{\eta}(s,t) \end{pmatrix}. \tag{9.29}$$

The matricial equations (9.28) and (9.29) can be rewritten in a reduced form, with the obvious meaning, as follows

$$\Omega_s = M \cdot \Omega, \qquad \Omega_t = P \cdot \Omega.$$

Then from the integrability condition $\Omega_{st} = \Omega_{ts}$ to find a solution, we obtain

$$M_t - P_s + [M, P] = 0,$$

which automatically implies that

$$\phi_t + \varphi_s = -i\,R\,\phi, \qquad R_s = \frac{1}{2}\,i(\phi\,\overline{\varphi} - \overline{\phi}\,\varphi). \tag{9.30}$$

Observe that in the above equations there are two undetermined functions, so that to identify them we ask the curve $\gamma(s)$ evolves in a suitable way, which we choose to be LIE (9.26). In other words, $\gamma(s)$ should evolve generating a Hasimoto surface. That equation can be written as

$$\alpha_t = \alpha_s \times \alpha_{ss} = T \times (\kappa N) = \kappa\,B,$$

so that this equation is also known as the binormal flow. Thus we obtain

$$T_t = \alpha_{st} = \alpha_{ts} = (\kappa\,B)_s = -\kappa\,\tau\,N + \kappa_s\,B. \tag{9.31}$$

On the other hand, a straightforward computation yields

$$\phi_s\,\overline{\eta} = \left[\kappa_s\,e^{i\int_0^s \tau(u,t)\,du} + i\kappa\,\tau\,e^{i\int_0^s \tau(u,t)\,du}\right]e^{-i\int_0^s \tau(u,t)\,du}(N - i\,B),$$

that is,

$$\phi_s\,\overline{\eta} = (\kappa_s + i\kappa\,\tau)(N - i\,B).$$

Similarly,

$$\overline{\phi}_s\,\eta = (\kappa_s - i\kappa\,\tau)(N + i\,B),$$

and now we combine both equations with (9.31) to have that

$$\phi_s\,\overline{\eta} - \overline{\phi}_s\,\eta = -2i\,T_t.$$

Finally, as $T_t = -\frac{1}{2}\,\overline{\varphi}\,\eta - \frac{1}{2}\,\varphi\,\overline{\eta}$ we get

$$\phi_s\,\overline{\eta} - \overline{\phi}_s\,\eta = -2i\left(-\frac{1}{2}\,\overline{\varphi}\,\eta - \frac{1}{2}\,\varphi\,\overline{\eta}\right) = i\varphi\,\overline{\eta} + i\overline{\varphi}\,\eta.$$

This equation allows us to identify the function $\varphi(s,t)$ as

$$\varphi(s,t) = -i\phi_s(s,t). \tag{9.32}$$

To determine the second function we bring the above equation to the compatibility condition (9.30) to obtain

$$R_s = \frac{1}{2}\, i\left(i\,\phi\,\overline{\phi}_s + i\,\overline{\phi}\phi_s\right) = -\frac{1}{2}\left(\phi\,\overline{\phi}_s + \overline{\phi}\phi_s\right).$$

Therefore,

$$R_s = -\frac{1}{2}\frac{\partial}{\partial s}|\phi|^2,$$

and

$$R(s,t) = -\frac{1}{2}\,|\phi(s,t)|^2 + A(t), \tag{9.33}$$

$A(t)$ being a function of t. Now, from (9.32) and (9.33), the first compatibility condition (9.30) can be written as follows

$$\phi_t - i\,\phi_{ss} = -i\left(-\frac{1}{2}|\phi|^2 + A\right)\phi,$$

so that

$$i\phi_t + \phi_{ss} + \frac{1}{2}(|\phi|^2 - 2A)\,\phi = 0,$$

is the cubic Schrödinger equation (9.27).

9.7 Elastic Curves Versus LIE

A natural problem is to find the filaments which, evolving under LIE, preserve its shape. In other words, filaments evolving over time without changing shape, only position. They are known as solitons or congruence solutions of LIE. That is equivalent, from a geometric viewpoint, to consider curves whose binormal flow is a one-parameter group of motions in $\mathbb{R}^3$.

Let us start with the following simple case. If a plane curve $\alpha(s)$ evolves in the direction of its binormal sweeping out a right cylinder, then it provides a filament (a solution of LIE) if and only if $\alpha(s)$ is a circle. In other words, circular right cylinders are the only right cylinders which are Hasimoto surfaces.

On the other hand, suppose that $\alpha(s) = (x(s), 0, z(s))$ lies in the plane $y = 0$ and we evolve it by a screw motion around the z-axis, so that it sweeps out a helicoid, that is,

$$\alpha(s,t) = R_t(\alpha(s)) = (x(s)\cos t, x(s)\sin t, z(s) + bt),$$

and then we ask for those plane curves whose helicoidal evolution provides a solution of (9.26). It is easy to see that the helicoids providing Hasimoto surfaces are actually revolution surfaces ($b = 0$) which are obtained just rotating curves whose curvature function is a multiple of the distance to some straight line parallel to the axis. Measuring it with respect to the own axis, we have $\kappa(s) = x(s)$.

Let us give some details on plane curves whose curvature function is a multiple of the distance to some straight line. We may assume, without loss of generality, that the distance to the y-axis is given by $\kappa(s) = \mu\, x(s)$, $\mu \in \mathbb{R}$, for a certain plane curve $\alpha(s) = (x(s), y(s))$. It is clear that the assumption can be written as

$$\begin{aligned} x''(s) + \mu\, x(s)\, y'(s) &= 0, \\ y''(s) - \mu\, x(s)\, x'(s) &= 0, \end{aligned}$$

so that the second equation can be integrated to get $y'(s) = \frac{\mu}{2} x^2(s) + A$ for some constant A. So, we put this information in the first equation to obtain

$$x''(s) + \frac{\mu^2}{2} x^3(s) + A\mu\, x(s) = 0$$

which in terms of the curvature gives

$$\kappa''(s) + \frac{1}{2}\kappa^3(s) + A\mu\,\kappa(s) = 0. \tag{9.34}$$

To prove the converse, let $\alpha(s)$ be a curve whose curvature function is a solution of (9.34) for $\mu \neq 0$. Then we define a new curve $\beta(s) = (x(s), y(s))$ where

$$\begin{aligned} x(s) &= \frac{1}{\mu}\kappa(s), \\ y(s) &= \int_0^s \left(\frac{\mu}{2} x^2(s) + A\right) du. \end{aligned}$$

A direct computation involving the Eq. (9.34) allows one to see that

$$\frac{d}{ds}|\beta'(s)|^2 = \frac{d}{ds}\left[(\kappa')^2 + \frac{1}{4}\kappa^4 + A\mu\kappa^2 + A^2\mu^2\right] = 2\kappa'\left(\kappa'' + \frac{1}{2}\kappa^3 + A\mu\kappa\right) = 0,$$

which shows that $\beta(s)$ is, up to some constant, parameterized by its arc-length. Moreover, it is not difficult to check that $\beta''(s) = \mu\, x(s) = (-y'(s), x'(s))$ which shows that the curvature of β satisfies

$$\kappa_\beta(s) = \mu\, x(s) = \kappa(s),$$

so both curves α and β are congruent.

Remark 9.7.1 It should be observed that the Eq. (9.34) is just the field equation, in the plane, associated to the classical Bernoulli's model of elastica (1.1) (see [69] and [70] and references therein). Therefore *the plane curves with curvature function being a multiple of the distance to some straight line are just the elastic curves in the plane*.

Consequently, we have the following result

A helicoid in $\mathbb{R}^3$ is a Hasimoto surface if and only if it is a revolution surface whose profile curve is an elastica in the Euclidean plane.

Even more, the solution to the general problem is known (see [63] and references therein) and now we are going to give some details.

In order to understand it, we use the tool provided by Killing vector field along a curve. Given a curve $\alpha(s)$ and a Killing field V in $\mathbb{R}^3$, we know that V generates a one-parameter group $\{\psi_t \,:\, t \in \mathbb{R}\}$ of motions in $\mathbb{R}^3$. We restrict V along $\alpha(s)$, writing $V(s)$, and take a variation of $\alpha(s)$ in the direction of the field to obtain

$$\alpha(s,t) = \psi_t(\alpha(s)).$$

It is clear that variation preserves all geometric properties of the curve. In particular, speed, curvature and torsion, so that we have (see (1.8), (1.9) and (1.11))

$$V(x) = 0, \qquad V(\kappa) = 0, \qquad V(\tau) = 0.$$

Let us now consider a curve $\alpha(s)$ in $\mathbb{R}^3$ and a vector field $V(s)$ along it such that $V(x) = V(\kappa) = V(\tau) = 0$. Then we will say that $V(s)$ is a Killing vector field along $\alpha(s)$. From the above discussion, the restriction of any Killing vector field, in $\mathbb{R}^3$, to a curve provides a Killing vector field along it. It is a pleasant surprise that the converse also holds (see [6]). More precisely, *any Killing vector field $V(s)$ along a curve $\alpha(s)$ in $\mathbb{R}^3$ is the restriction to the curve of a Killing vector field V in $\mathbb{R}^3$*. To see it, observe that the Killing vector fields $V(s)$ along a curve $\alpha(s)$ are the solutions of the linear system of differential equations $V(x) = V(\kappa) = V(\tau) = 0$. From (1.8) , (1.9) and (1.11), the set of solutions is a 6-dimensional vector space, which is just the dimension of both the isometry group of $\mathbb{R}^3$ and that of Killing vector fields on $\mathbb{R}^3$. As the restriction of any Killing vector field on $\mathbb{R}^3$ to a curve is a Killing vector field along it, the extension is guaranteed.

Coming back to the problem, we wish to characterize the curves $\alpha(s)$ in $\mathbb{R}^3$ such that

$$V(x) = 0, \qquad V(\kappa) = 0, \qquad V(\tau) = 0, \qquad V(s) = \kappa(s)\, B(s).$$

From (1.8) we obtain

$$V(x) = \big[\kappa'\langle B, T\rangle - \kappa\tau\langle N, T\rangle\big]\, x = 0,$$

which gives no information. Now we compute

$$\nabla_T\,(\kappa\, B) = -\kappa\tau\, N + \kappa'\, B,$$
$$\nabla_T^2\,(\kappa\, B) = \kappa^2\tau\, T - (2\kappa'\tau + \kappa\tau)N + (\kappa'' - \kappa\tau^2)B,$$

and the use (1.9) to get

$$(\kappa\, B)(\kappa) = \langle \nabla_T^2\,(\kappa\, B), N\rangle = -\frac{d}{ds}(\kappa^2\tau).$$

Therefore, this variation vanishes if and only if

$$\kappa^2\,\tau = c, \tag{9.35}$$

c being a constant. Finally, from (1.11) we deduce that

$$(\kappa\, B)(\tau) = -\frac{d}{ds}\left(\frac{1}{\kappa}(\kappa'' - \kappa\tau^2\right) - \kappa\kappa',$$

so that $(\kappa\, B)(\tau) = 0$ means that

$$\frac{1}{\kappa}(\kappa'' - \kappa\tau^2 + \frac{1}{2}\kappa^2 = A,$$

A being a constant. Putting all together we have

$$\kappa'' - \kappa\tau^2 + \frac{1}{2}\kappa^3 - \frac{\lambda}{2}\kappa = 0, \qquad A = \frac{\lambda}{2}. \tag{9.36}$$

Now we have to realize that Eqs. (9.35) and (9.36) are just the Euler-Lagrange (9.18) and (9.19) of elastic curves in $\mathbb{R}^3$. So we have shown the following result

A curve $\alpha(s)$ in $\mathbb{R}^3$ is the initial condition of a soliton of LIE if and only if $\alpha(s)$ is an elastic curve. In other words, the filaments evolving under LIE, without changing shape, only position, are generated by elastic curves in $\mathbb{R}^3$.

This result, regardless of its beauty and elegance, provides an a priori unsuspected connection between the theory of elasticity and fluid dynamics.

Chapter 10
Variational Models

The notion of curvature of a curve can be given in any Riemann space. Indeed, when we have a regular curve $\gamma(t)$ in a Riemannian space $(M^n, g = \langle , \rangle)$, we can define, as in the Euclidean space, the arc length

$$s(t) = \int_{t_o}^{t} ||\gamma'(u)||\, du.$$

Then the curve can be arc lenth reparameterized so that we get a unit tangent vector, with respect to the new parameter

$$T(s) = \gamma'(s) = \frac{d\gamma(s)}{ds}, \qquad ||T(s)|| = 1.$$

Then, using the Riemannian connection ∇, associated to the metric, we can define the **geodesic curvature** at any point of $\gamma(t)$ as follows,

$$\kappa(s) = \|\nabla_T T\|(s).$$

It is clear that this definition can be done independently of the dimension of the space and it is always non-negative. However, when the dimension is two, as in the case of the plane, a subtle definition of geodesic curvature can be made.

Take an orientation of the surface and rotate ninety degrees T in the positive direction to obtain the unit normal $N = J(T)$ and define

$$\kappa(s) = \langle \nabla_T T, N \rangle (s).$$

A. Ferrandez Izquierdo, *Helices*, RSME Springer Series 16,
https://doi.org/10.1007/978-3-032-09663-0_10

From a geometric point of view, a natural and interesting question arises as follows:

Let $(M^n, \langle, \rangle)$ be an n-dimensional Riemannian manifold. On a certain space of clamped curves, C, consider the functional that measures the total curvature:

$$\mathcal{F} : \mathbf{C} \to \mathbb{R}, \quad \mathcal{F}(\alpha) = \int_\alpha \kappa(s)\, ds.$$

Problem: Study the Variational Problem $(\mathbf{C}, \mathcal{F})$ in $(M^n, \langle, \rangle)$
As we have seen before, the problem turns out to be trivial when $(M^n, \langle, \rangle)$ is a Euclidean plane, at least in the cases where **C** is the space of closed or clamped curves. Then the functional is constant on each homotopy class of curves.

10.1 Particle Models

This problem is of great geometrical and physical interest, because it is closely related to particle theory. Indeed, the search for Lagrangians that allow us to describe particles with, or without, mass and spin has been (and continues to be) a problem and a chimera with a long history.

The treatment that has traditionally been given to this problem is based on a very simple idea and has become a routine: **expand the original spacetime by adding auxiliary coordinates that in some sense represent, or describe, the spinor degrees of freedom.**

It is clear that the method, from a mathematical point of view, is not too smart. For instance, in many cases, the enlarged space is not very well known. Even from a physical point of view, the classic procedure presents, from its origin, some problems such as:

(1) It gives up the old aspiration to get an intrinsic description of the particles, that is, in the spacetime where they live.
(2) It gives up obtaining models that describe, of a consistent way, particles with own maximal acceleration.
(3) The quantization of these models is practically impossible, which means that this strate cannot lead us to find supersymmetric particle models.

Therefore, since the classical procedure presents serious difficulties, new ideas must be found in searching for Lagrangians that provide particles models.

10.2 Rigidity Models

At the beginning of the nineties, of the last century, arose a new way to study particles. It seems that the cause of its birth has a lot to do with a certain mimicry towards the, already known by then, **bosonic string theory** of Nambu-Goto-Polyakov-Kleinert and that we will see later.

At the origin of this new procedure there are, essentially, important differences, with respect to the classic treatment:

(1) Models are intrinsic. That is, they are set to describe the particles in the original spacetime where the dynamics of the system take place. No additional degrees of freedom are added.
(2) The classical Lagrangian, mimicking that of the free falling particle in a strange space where it was presumably unified gravity with another interaction, is replaced by another one rather more **complicated** and **subtle**.
(3) The complication comes from the inclusion of derivatives of order greater than one in the Lagrangian density (the classical model includes only derivatives of order one, specifically the length).
(4) The subtlety arises from the Lagrangian density, which:
    (a) Cannot be randomly built with derivatives of higher orders.
    (b) Must be invariant under the motions of the space.
    (c) Must be a function of the curvatures of the trajectories of the particles.

Therefore, **the beauty of the model lies in the idea of considering that spinor degrees of freedom are encoded in the geometry of the trajectories of the particles**

In a more concrete way, first take a Riemannian space $(M^n, \langle , \rangle)$. In a second step we will consider Lorentz spaces with the corresponding technical differences in the results. Then each curve $\gamma(s)$, sufficiently regular and parameterized by its arc length, has $n-1$ curvature functions that naturally, together with the arc length, provide n geometric invariants (which are invariant by isometries or motions of the ambient space). These curvatures are defined, as in the case of Euclidean space, using the Riemannian connection associated with the metric of the space. Indeed, the first curvature, $\kappa_1(s)$, is just the function that we have already defined, that is, we have the following first Frenet equation

$$\nabla_T T(s) = \kappa_1(s)\, N(s).$$

Now taking into account that $\langle \nabla_T T, N \rangle + \langle T, \nabla_T N \rangle = 0$, we can compute the derivative of the normal N to get

$$\nabla_T N(s) = -\kappa_1(s)\, T(s) + \kappa_2(s)\, \xi_3(s),$$

where $\xi_3(s)$ is a unit vector field, along the curve, orthogonal to both $T(s)$ and $N(s)$. Taking again derivative we find

$$\nabla_T \, \xi_3(s) = -\kappa_2(s)\, N(s) + \kappa_3(s)\, \xi_4(s),$$

where $\xi_4(s)$ is a unit vector field, along the curve, orthogonal to $\{T(s), N(s), \xi_3\}$, where $\kappa_3(s)$ is the third curvature function. Of course, you can continue until you get an orthonormal frame $\{T, N, \xi_3, \ldots, \xi_n\}$ along the curve, whose variation is governed by the curvature functions $\kappa_1, \kappa_2, \ldots, \kappa_{n-1}$ according to the following Frenet equations

$$\begin{aligned}
\nabla_T \, T(s) &= \kappa_1(s)\, N(s) \\
\nabla_T \, N(s) &= -\kappa_1(s)\, T(s) + \kappa_2(s)\, \xi_3(s) \\
\nabla_T \, \xi_3(s) &= -\kappa_2(s)\, N(s) + \kappa_3(s)\, \xi_4(s) \\
&\;\;\vdots \\
\nabla_T \, \xi_{n-1}(s) &= -\kappa_{n-2}(s)\, \xi_{n-2}(s) + \kappa_{n-1}(s)\, \xi_n(s) \\
\nabla_T \, \xi_n(s) &= -\kappa_{n-1}(s)\, \xi_{n-1}(s).
\end{aligned}$$

So, the Lagrangian densities in the new models must be functions $\Omega(\kappa_1, \kappa_2, \cdots, \kappa_{n-1})$ of the above curvatures.

Given a Lagrangian density, consider a certain space of curves, $\mathcal{C}$, on $(M^n, \langle , \rangle)$, the functional $\mathcal{F} : \mathcal{C} \to \mathbb{R}$ and the associated variational problem

$$\mathcal{F}(\alpha) = \int_\alpha \Omega\,(\kappa_1, \kappa_2, \cdots, \kappa_{n-1})\,(s)\, ds.$$

The trajectories of the particles in such a model are the critical curves of that variational problem.

10.3 Some Examples

(1) **Free falling particles.** It seems clear that the simplest choice of Lagrangian density in the above context is

$$\Omega\,(\kappa_1, \kappa_2, \cdots, \kappa_{n-1}) = C, \text{ constant.}$$

In this case the model describes the free falling particles, only submitted to gravity action in $(M^n, \langle , \rangle)$. It is well known that the trajectories are geodesics of $(M^n, \langle , \rangle)$.

(2) **Non-massive bosons.** In 1990 M. S. Plyushchay [87] proposed a beautiful Lagrangian to describe the non-massive bosons.

The Lagrangian density governing this model is, up to a coupling constant, the acceleration of the particle. That is, the curvature of the trajectories

$$\Omega(\kappa_1, \kappa_2, \cdots, \kappa_{n-1}) = C\,\kappa_1.$$

The Lagrangian proposed by Plyushchay to describe non-massive bosons measures, up to a constant, the total curvature of the trajectories

$$\mathcal{F}(\alpha) = C \int_\alpha \kappa_1(s)\, ds.$$

This is, up to an innocent constant, the variational problem that we found interesting from a geometric viewpoint.

(3) Massive bosons. The above model extends using a Lagrange multiplier to study massive particles. We then have the following Lagrangian density

$$\Omega(\kappa_1, \kappa_2, \cdots, \kappa_{n-1}) = C\,\kappa_1 + m.$$

In this case, the Lagrangian that governs the model, proposed by Plyushchay to describe massive bosons, is

$$\mathcal{F}(\alpha) = \int_\alpha (C\,\kappa_1(s) + m)\; ds.$$

(4) Models with tachyonic states. A model for relativistic particles governed by a Lagrangian density linear in torsion (second curvature) appears in the mechanism of Bose-Fermi transmutation. Both classic and quantum point of view have been systematically studied in a number of articles such as [80, 88] and [89].

In these models, the Lagrangian density is of the form

$$\Omega(\kappa_1, \kappa_2, \cdots, \kappa_{n-1}) = C\,\kappa_2,$$

and the action that governs the dynamics of the model is then

$$\mathcal{F}(\alpha) = \int_\alpha (C\,\kappa_2(s) + m)\; ds.$$

In short, to describe a system of relativistic particles the following items are required:

(a) A spacetime, $(M, \langle , \rangle)$, where particles will live and evolve.
(b) A space of curves, $\mathcal{C}$, in the spacetime, which would play the role of dynamic variables of the theory.
(c) An admissible action, $\mathcal{F} : \mathcal{C} \to \mathbb{R}$, which will govern the theory.

(d) Some trajectories, the particles of the system $[(M, \langle,\rangle), \mathcal{C}, \mathcal{F}]$ which evolve through curves that are critical points of $\mathcal{F} : \mathcal{C} \to \mathbb{R}$.

10.4 Bosons

In an $(n \geq 2)$-dimensional Riemannian space $(M, \langle,\rangle)$ take two points $p, q \in M$ and the space of curves Λ joining them

$$\Lambda = \{\alpha : [a, b] \to M \, : \, \alpha(a) = p, \; \alpha(b) = q\}.$$

On the other hand, consider in Λ new curves $\Gamma : (-\varepsilon, \varepsilon) \to \Lambda$ such that $\Gamma(v) \in \Lambda$, $v \in (-\varepsilon, \varepsilon)$. We will say that this curve is differentiable when it is so as seen in M. In other words, from the curve $\Gamma : (-\varepsilon, \varepsilon) \to \Lambda$ we build a map, also denoted by Γ, as follows

$$\Gamma : [a, b] \times (-\varepsilon, \varepsilon) \to M, \qquad \Gamma(t, v) = \Gamma(v)(t).$$

We will say that the curve Γ in Λ is differentiable when it is so as a map on M. In these conditions we can define two vector fields along the map

$$X(t, v) = \frac{\partial \Gamma}{\partial t}(t, v), \qquad V(t, v) = \frac{\partial \Gamma}{\partial v}(t, v),$$

which we will call, respectively, longitudinal and transversal.

Imagine now a curve $\Gamma : (-\varepsilon, \varepsilon) \to \Lambda$, passing through a point of Λ, i.e., $\Gamma(0) = \alpha \in \Lambda$, and we wish to compute its tangent vector $\Gamma'(0)$ as

$$\Gamma'(0) = \frac{\partial \Gamma}{\partial v}(t, 0) = V(t, 0).$$

Furthermore, as the curves in Λ connect two fixed points of M, we have that $V(a, 0) = V(b, 0) = 0$. We then obtain that *any tangent vector to Λ at a point α is seen, in M, as a vector field $V(t) = V(t, 0)$, along the curve $\alpha(t)$, vanishing at the ending points.*

The converse also holds. Given a vector field $V(t)$ along a curve $\alpha \in \Lambda$, vanishing at the ending points, $V(a) = V(b) = 0$, we can define a curve in Λ, passing through α, with speed vector $V(t)$. Indeed,

$$\Gamma : (-\varepsilon, \varepsilon) \to \Lambda, \qquad \Gamma(v)(t) = \exp_{\alpha(t)}(v\, V(t)).$$

It is clear that

$$\Gamma(0) = \alpha, \qquad \Gamma'(0) = V.$$

As a consequence, *the tangent space to* Λ *at a point can be seen in* M *as the space of vector fields along the curve* $\alpha(t)$ *vanishing at the ending points*

$$T_\alpha \Lambda = \{V(t) \in T_{\alpha(t)}M \ : \ V(a) = V(b) = 0\}.$$

Let us now formulate the variational problem. For any real number m, consider the functional $\mathcal{F}_m : \mathcal{C} \to \mathbb{R}$ defined by

$$\mathcal{F}_m(\alpha) = \int_\alpha (\kappa(s) + m) ds.$$

We are going tu study the variational problem, in particular the solutions of the field, or Euler-Lagrange, equations associated to this functional. In other words, the trajectories of the particles -bosons- associated to these models. To compute the Euler-Lagrange equations associated to these functionals we have to know the differential as a linear map

$$\delta\mathcal{F}_m(\alpha) : T_\alpha \Lambda \to \mathbb{R}.$$

Given $V \in T_\alpha \Lambda$, we have to compute $\delta\mathcal{F}_m(\alpha)[V]$. To do that we consider a variation of the curve α in the direction of V. Set

$$\Gamma : [a, b] \times (-\varepsilon, \varepsilon) \to M, \quad \Gamma(t, 0) = \alpha(t), \quad \frac{\partial \Gamma}{\partial v}(t, 0) = V(t).$$

Defining as above the longitudinal $X(t, v)$ and transversal $V(t, v)$ vector fields, we get

$$[X, V] = 0.$$

On the other hand, given the function measuring the speed $x(t, v) = \|X(t, v)\|$ of the longitudinal curves, we obtain the unit longitudinal vector field

$$T(t, v) = \frac{X(t, v)}{x(t, v)}, \qquad X(t, v) = x(t, v)\, T(t, v).$$

Bringing this information to the bracket we find that

$$\nabla_{x\,T}\, V = \nabla_V\,(x\,T) \quad \Longrightarrow [V, T] = -\frac{V(x)}{x}\, T.$$

We wish to compute

$$\delta\mathcal{F}_m(\alpha)[V] = \left\{\frac{\partial}{\partial\, v}\, [\mathcal{F}_m(\Gamma(t, v))]\right\}_{v=0} = \left\{\frac{\partial}{\partial\, v}\left[\int_a^b (\kappa(t, v) + m)\, x(t, v)\, dt\right]\right\}_{v=0}.$$

To do so we need some formulas to measure the transversal variations of the two functions, speed and curvature of the longitudinal curves, appearing in the above integrand. In a first step we have

$$\frac{\partial x}{\partial v} = \frac{\partial}{\partial v}\sqrt{\langle X, X\rangle} = \frac{\langle \nabla_V X, X\rangle}{x} = \frac{\langle \nabla_X V, X\rangle}{x} = \langle \nabla_T V, T\rangle\, x.$$

To compute the transversal variation of the curvature function of the longitudinal curves, we have

$$\frac{\partial \kappa^2}{\partial v} = V\langle \nabla_T T, \nabla_T T\rangle = 2\langle \nabla_V \nabla_T T, \nabla_T T\rangle.$$

On the other hand, from the formula of the curvature, we get

$$\nabla_V \nabla_T T = \nabla_T \nabla_V T + \nabla_{[V,T]} T + R(V,T)T,$$

so that

$$\begin{aligned}\frac{\partial \kappa^2}{\partial v} &= 2\langle \nabla_T(\nabla_T V + [V,T]) - \frac{V(x)}{x}\nabla_T T + R(V,T)T, \nabla_T T\rangle \\ &= 2\langle \nabla_T^2 V - T\left(\frac{V(x)}{x}\right) T - 2\frac{V(x)}{x}\nabla_T T + R(V,T)T, \nabla_T T\rangle.\end{aligned}$$

Simplifying we find

$$2\kappa\frac{\partial \kappa}{\partial v} = 2\kappa\langle \nabla_T^2 V, N\rangle - 4\frac{V(x)}{x}\kappa^2 + 2\kappa\langle R(V,T)T, N\rangle,$$

arriving at

$$\boxed{\frac{\partial \kappa}{\partial v} = \langle \nabla_T^2 V, N\rangle - 2\frac{V(x)}{x}\kappa + \langle R(V,T)T, N\rangle}$$

We use these formulas below and, repeatedly, the integration by parts to compute

$$\begin{aligned}\delta\mathcal{F}_m(\alpha)[V] &= \left\{\frac{\partial}{\partial v}\left[\int_{\Gamma(t,v)} (\kappa(t,v) + m)\, ds\right]\right\}_{v=0} \\ &= \left\{\frac{\partial}{\partial v}\left[\int_a^b (\kappa(t,v) + m)\, x(t,v)\, dt\right]\right\}_{v=0} \\ &= \int_a^b \frac{\partial \kappa}{\partial v}\, x\, dt + \int_a^b (\kappa + m)\, \frac{\partial x}{\partial v}\, dt\end{aligned}$$

$$= \int_\alpha \frac{\partial \kappa}{\partial v} ds + \int_a^b (\kappa + m)\langle \nabla_T V, T\rangle x\, dt$$

$$= \int_\alpha \frac{\partial \kappa}{\partial v} ds + \int_\alpha (\kappa + m)\langle \nabla_T V, T\rangle ds$$

$$= \int_\alpha \langle \nabla_T \nabla_T V, N\rangle ds - 2\int_\alpha \langle \nabla_T V, T\rangle \kappa\, ds$$

$$+ \int_\alpha \langle R(V, T)T, N\rangle ds + \int_\alpha \langle \nabla_T V, T\rangle \kappa\, ds + m\int_\alpha \langle \nabla_T V, T\rangle ds$$

$$= \int_\alpha \frac{d}{ds}\langle \nabla_T V, N\rangle ds - \int_\alpha \langle \nabla_T V, \nabla_T N\rangle ds - \int_\alpha \frac{d}{ds}\langle V, T\rangle \kappa\, ds$$

$$+ \int_\alpha \langle V, \nabla_T T\rangle \kappa\, ds + \int_\alpha \langle R(V, T)T, N\rangle ds + m\int_\alpha \frac{d}{ds}\langle V, T\rangle ds$$

$$- m\int_\alpha \langle V, \nabla_T T\rangle ds.$$

There are still two last integrations by parts corresponding to the second and third summands. In the following expression we will group the summands corresponding to exact integrals

$$\delta\mathcal{F}_m(\alpha)[V] = \int_\alpha \frac{d}{ds}\langle \nabla_T V, N\rangle ds - \int_\alpha \frac{d}{ds}\langle V, \nabla_T N\rangle ds - \int_\alpha \frac{d}{ds}(\kappa\langle V, T\rangle) ds$$

$$+ m\int_\alpha \frac{d}{ds}\langle V, T\rangle ds + \int_\alpha \langle V, \nabla_T^2 N\rangle ds + \int_\alpha \kappa'\langle V, T\rangle ds$$

$$+ \int_\alpha (\kappa - m)\langle V, \nabla_T T\rangle ds + \int_\alpha \langle R(V, T)T, N\rangle ds.$$

We then obtain the following formula for the differential of the functional

$$\boxed{\delta\mathcal{F}_m(\alpha)[V] = \int_\alpha \langle \Omega(\alpha), V\rangle ds + [\mathcal{B}(\alpha, V)]_a^b}$$

where $\Omega(\alpha)$ and $\mathcal{B}(\alpha, W)$ represent the Euler-Lagrange and boundary operators, respectively, which in this case they are given by

$$\boxed{\Omega(\alpha) = \nabla_T^2 N + \kappa' T + (\kappa - m)\nabla_T T + R(N, T)T}$$

and

$$\boxed{\mathcal{B}(\alpha, V) = \langle(\nabla_T V, N\rangle + \langle(V, (m - \kappa)T - \nabla_T N\rangle}$$

Of course, one can use the Frenet equations of α in $(M, \langle\, , \rangle)$ to get

$$\Omega(\alpha) = -(\tau^2 + m\kappa)N - \tau_s B - \tau\eta + R(N, T)T,$$

where B is the unit binormal, τ the torsion and τ_s its derivative with respect to the arc length parameter. Finally, η is a vector field along α which is orthogonal to the subbundle $\{T, N, B\}$.

As we are working in Λ, all longitudinal curves start and end at the same points, so that $V(a) = V(b) = 0$. Then, the boundary operator is

$$[\mathcal{B}(\alpha, V)]_a^b = \langle(\nabla_T V(b), N(b)\rangle - \langle(\nabla_T V(a), N(a)\rangle.$$

To get the Euler-Lagrange equations we consider first order boundary conditions. Therefore, our space of curves Λ will be formed by the so called *clamped curves in two points*. More precisely, take two points $p, q \in M$ and two tangent vectors $\vec{u} \in T_p M$, $\vec{v} \in T_q M$, and the space of curves Λ connecting the points and tangent to the fixed vectors

$$\alpha(a) = p, \quad \alpha(b) = q, \quad \alpha'(a) = \vec{u}, \quad \alpha'(b) = \vec{v}.$$

Having this boundary conditions and taking variations of the clamped curves we obtain

$$[V, T] = -\frac{V(x)}{x} T \quad \Longrightarrow \quad \nabla_T V = \partial_v(\log x)T + \nabla_V T.$$

This formula shows that if the curves of the variation are tangent at the ending points, then $\nabla_V T(0) = \nabla_V T(L) = 0$, so that

$$[\mathcal{B}(\gamma, V)]_a^b = 0.$$

As a conclusion, the critical points of the variational problem are the curves $\alpha \in \mathcal{C}$ such that

$$\delta\mathcal{F}_m(\alpha)[V] = 0, \qquad V \in T_\alpha\Lambda.$$

Therefore, they are the solutions of the following field,or Euler-Lagrange, equation

$$\boxed{R(N, T)T = (\tau^2 + m\kappa)N + \tau_s B + \tau\eta}$$

As a consequence, the trajectories of the model, that is, its dynamics, work on curves that are solutions of the above equation. There the ambient space has a key presence through its curvature operator, essentially acting on the osculating plane of the trajectories. The solution of this equation is then encoded in the degree of knowledge and control that we have over the curvature of the ambient space.

There are two interesting contexts where we can obviously progress in the knowledge of trajectories: (1) when the ambient space is a **surface**, that is, it has dimension two; and (2) when the ambient space, of any dimension, has **constant curvature**.

10.5 Bosons Dynamics on a Surface

Let us take a 2-dimensional space and denote by G its curvature function. Then, the above field equation can be written as

$$\boxed{G = m\kappa}$$

where it is understood that, in the first member of the equation, the Gaussian curvature is constrained on the trajectories.

Let us see some examples:

(i) **Flat surfaces.** When the Gaussian curvature vanishes, $G = 0$, then $m\,\kappa = 0$. Therefore, the dynamics of the non-massive boson is trivial, i.e., any curve of the surface becomes a possible trajectory. However, the massive boson dynamics works on geodesics of the surface.

(ii) **Non-zero constant curvature surfaces.** In this case the dynamics of the non-massive boson can not hold, while that of the massive one works on curves with constant curvature G/m.

(iii) **Non-massive bosons.** It is clear that on any surface, the dynamics of non-massive bosons ($m = 0$) is produced through curves on which the Gaussian curvature of the surface vanishes, in other words, the trajectories are curves of parabolic points. For example, consider a curve, parameterized by the arc length, in the open half plane $y = 0$ and $z > 0$,

$$\gamma(s) = (f(s), 0, h(s)), \quad f(s) > 0, \quad (f'(s))^2 + (h'(s))^2 = 1, \quad s \in \mathrm{I} \subset \mathbb{R}.$$

Then it can be rotated around the z-axis to obtain a revolution surface $\mathbf{S}_\gamma$ that can be written as

$$X(s, v) = (f(s)\,\cos v,\, f(s)\,\sin v,\, h(s)), \qquad s \in \mathrm{I}, \quad v \in [0, 2\pi],$$

whose Gaussian curvature is given by

$$G(s, v) = -\frac{f''(s)}{f(s)}.$$

So the set of parabolic points is formed by the parallels corresponding to the zeros of f''. But it is observed that the normal to the curve generatrix of the surface is $(-h''(s), 0, f''(s))$, so these parallels are those having a tangent perpendicular to the axis of rotation (the z-axis).

(iv) **The charge variational problem.** It often happens that some questions, at a first sight in principle without apparent relationship between them, are closely related. Here is a good example. Let us consider a surface $(M^2, g = \langle\, ,\rangle)$ isometrically immersed in $\mathbb{R}^3$. If it is considered orientable, then we have, globally defined, its Gauss map

$$N : M \to \mathbb{S}^2(1), \qquad N(p) \perp T_pM, \quad \langle N(p), N(p)\rangle = 1.$$

Let us denote by $d\sigma^2$ the area element on the unit sphere, so that $N^*(d\sigma^2)$ defines what in soliton theories is called the *charge density on M*. It is known that

$$N^*(d\sigma^2) = K\,dA,$$

where dA represents the surface area element.
Let us now take a bounded domain $\Omega \subset \mathbb{R}^2$ whose boundary $\partial\Omega$ is conveniently differentiable. Next, consider the embeddings space of the domain in the surface

$$\mathcal{E}(\Omega, M) = \{\phi : \Omega \to M \,:\, \phi \text{ is embedding}\}.$$

Consider now the total charge action defined as follows

$$\mathfrak{C} : \mathcal{E}(\Omega, M) \to \mathbb{R}, \qquad \mathfrak{C}(\phi) = \int_{\phi(\Omega)} G\,dA.$$

It is clear that the maximum or minimum problem associated with this functional on a round sphere, $\mathbb{S}^2(r)$, has no solution. On the other hand, taken a revolution torus $T(R, r)$ in $\mathbb{R}^3$ and Ω is considered to be an annulus, then the total charge reaches both its maximum and minimum in the two embeddings whose boundaries are formed, precisely, by the two parallels of parabolic points. Therefore, at least in the case of the revolution torus, the maximum and the minimum of the total charge are reached on embeddings whose boundaries are trajectories of non-massive bosons. But is this a general situation? The answer is yes, and even more general. The explanation is found in the following Gauss-Bonnet formula

$$\mathfrak{C}(\phi) + \mathcal{F}_0(\phi(\partial\Omega)) = 2\pi(1 - h),$$

where h denotes the number of holes of the domain Ω. The formula proves that, at least at the level of critical points, both problems are equivalent. A domain is critical for the total charge if and only if the charge vanishes on its boundary. It can also be proved that the stability of the critical points, for the total charge, is also reflected on the boundary. A critical domain is stable if and only if its boundary (which is a level curve for the charge) is lying in the set of regular points for the charge. This whole story can be seen as a beautiful holographic principle.

(v) **The Beem and Buseman surfaces.** We start by considering a warped product of the following type

$$\mathbf{S} = I \times_f N,$$

where I is a real interval, which can be the whole real line, f is a positive definite function on I, N is 1-dimensional manifold, for example $\mathbb{R}$ or $\mathbb{S}^1$, and evidently the metric is

$$g = dt^2 + f^2 d\theta^2.$$

For each $t \in I$, the curve $N_t = \{t\} \times N$ has constant curvature, $\kappa = -\frac{f'}{f}$, in $(\mathbf{S}, g)$. Let T be the unit tangent of N_t, the unit normal is ∂_t and the curvature operator is

$$R(\partial_t, T)T = -\frac{f''}{f}\partial_t.$$

Therefore, N_t is a solution of the field equation associated to $\mathcal{F}_m$ if and only if $f'' = m\, f'$.

In particular, **all curves N_t are solutions of the field equation associated to $\mathcal{F}_m$ if and only if $f : I \to \mathbb{R}$ is defined by $f(t) = A\, e^{mt} + B$, for constants** A and B.

This class of surfaces includes the Beem and Buseman surfaces. Therefore, these surfaces admit a foliation by solutions of the field equation for some $\mathcal{F}_m$.

Chapter 11
Bosons in Spaces Forms

As we know, the dynamics of bosons, in a Riemann space $(M^n, g = \langle , \rangle)$ with curvature R, is described by the following Euler-Lagrange equation

$$R(N, T)T = (\tau^2 + m\kappa)N + \tau_s B + \tau\eta.$$

When the space has constant curvature, C, the above equation is equivalent to the system

$$\begin{aligned} \tau^2 + m\kappa &= C, \\ \tau_s &= 0, \\ \tau\eta &= 0. \end{aligned}$$

The following information is immediately obtained:

(1) From the third equation we have that $\eta = 0$, which proves that the dynamics actually takes place in spaces of dimension $n \leq 3$. That is, up to topology, the trajectories are in $\mathbb{R}^3$ (Euclidean space), $\mathbb{S}^3$ (sphere) or $\mathbb{H}^3$ (hyperbolic space), depending on $C = 0$, $C > 0$ or $C < 0$, respectively.
(2) Also, from the second equation, it is seen that the paths must have constant torsion;
(3) Finally, the first equation leads to:

    (3a) **Massive bosons in spaces of constant curvature evolve through helices in three-dimensional spaces;**
    (3b) **Non-massive bosons in spaces of constant curvature evolve on paths with constant torsion in a Euclidean plane (when the torsion vanishes) or in a sphere of dimension three (when the torsion does not vanish and its square is the curvature of the sphere).**

A. Ferrandez Izquierdo, *Helices*, RSME Springer Series 16,
https://doi.org/10.1007/978-3-032-09663-0_11

11.1 Non-massive Bosons

When $m = 0$, the dynamics of bosons works on curves with constant torsion, in spaces of dimension $n \leq 3$ and satisfying

$$\tau^2 = C.$$

There are two possibilities: (i) Euclidean plane case, where the model is known to be trivial, since the action is constant over each homotopy class of curves; and (ii) The second case corresponds to trajectories in a space that, up to topology, is the sphere of dimension three. The problem then reduces to finding, in $\mathbb{S}^3(r)$, the curves with constant torsion given by $\tau = \pm\frac{1}{r}$.

11.2 Massive Bosons

When $m \neq 0$, the dynamics reduces to finding helices (constant curvature and torsion curves) in $M^3(C)$ (space of constant curvature C and dimension three) whose curvature and torsion satisfy

$$\tau^2 + m\kappa = C.$$

In any case, you get a one-parameter family of helices. Thus, in the case of a flat space, for example $\mathbb{R}^3$, one finds circular helices with pitch b located on circular cylinders with radius a, satisfying the condition

$$b^2 + ma(a^2 + b^2) = 0.$$

11.3 Bosons in $\mathbb{S}^3$

As we have just seen, the Riemannian space where the boson dynamics has the greatest richness in the three-dimensional sphere $\mathbb{S}^3(r)$. If the bosons are non-massive, their trajectories are curves with constant torsion equals, up to the sign, to the inverse of the radius. In the case of massive bosons, the trajectories are helices with curvature and torsion satisfying the following relation

$$\tau^2 + m\kappa = \frac{1}{r^2}.$$

Naturally, in both cases we want to know more about the trajectories. To do that, we will study some details of the delicious geometry of a three-dimensional sphere.

Without losing generality and for simplicity, we will consider the three-dimensional sphere $\mathbb{S}^3$ of radius one. We see it into $\mathbb{C}^2$, with the metric induced by the Euclidean one, as

$$\mathbb{S}^3 = \{z = (z_1, z_2) \in \mathbb{C}^2 \ : \ |z_1|^2 + |z_2|^2 = 1\}.$$

On the other hand, we consider the circle seen as the multiplicative group of the unit complex numbers

$$\mathbb{S}^1 = \{e^{it} \ : t \in \mathbb{R}\},$$

and a natural action is defined as follows (see Sect. 2.2):

$$\mathbb{S}^1 \times \mathbb{S}^3 \to \mathbb{S}^3, \qquad e^{it}\, z = \left(e^{it}\, z_1, e^{it}\, z_2\right).$$

For each point $z \in \mathbb{S}^3$ we have the corresponding orbit $\alpha_z(t) = e^{it}\, z$, whose tangent vector, at z, is $\alpha_z'(0) = i\, z$. That means that the orbit flow corresponds to the following vector field over $\mathbb{S}^3$ (the orbits appear as the integral curves of the following field)

$$V(z) = i\, z, \quad \text{for any } z \in \mathbb{S}^3,$$

which, evidently, is unitary and therefore has no zeros. We then have the orbit space, which can be represented as $\mathbb{S}^3/\mathbb{S}^1$. Our immediate purpose is to prove that this orbit space can be identified with a 2-dimensional sphere. For this we consider the Hopf map

$$\Pi : \mathbb{S}^3 \to \mathbb{S}^2, \quad \Pi(z_1, z_2) = \left(z_1\, \overline{z}_2, \frac{1}{2}(|z_1|^2 - |z_2|^2)\right),$$

where we are seeing the 2-dimensional round sphere in $\mathbb{C} \times \mathbb{R} \equiv \mathbb{R}^3$ with radius $1/2$. Then you have:

Two points of $\mathbb{S}^3$ are in the same orbit if and only if they have the same image by the Hopf map.

Proof Suppose that two points of $\mathbb{S}^3$ are in the same orbit, then

$$w = e^{it}\, z \quad \Rightarrow \quad w_j = e^{it}\, z_j, \ 1 \le j \le 2,$$

so

$$w_1\, \overline{w}_2 = e^{it}\, z_1\, e^{-it}\, \overline{z}_2 = z_1\, \overline{z}_2.$$

Furthermore, it is clear that

$$|w_1|^2 - |w_2|^2 = |z_1|^2 - |z_2|^2,$$

which shows the necessary condition. As for the converse, we suppose now that two points of $\mathbb{S}^3$ have the same image by the Hopf map, that is, $\Pi(z) = \Pi(w)$. You have then

$$\begin{aligned} w_1\,\overline{w}_2 &= z_1\,\overline{z}_2, \\ |w_1|^2 - |w_2|^2 &= |z_1|^2 - |z_2|^2, \\ |w_1|^2 + |w_2|^2 &= |z_1|^2 + |z_2|^2 = 1. \end{aligned}$$

From the last two equations it is obtained that $|w_1| = |z_1|$ and $|w_2| = |z_2|$, so two real numbers appear, θ_1 and θ_2, such that

$$w_1 = e^{i\theta_1}\, z_1, \qquad w_2 = e^{i\theta_2}\, z_2.$$

Finally, writing this information in the first of the three above equations, we arrive at

$$w_1\,\overline{w}_2 = e^{i(\theta_1-\theta_2)}\, z_1\,\overline{z}_2 = z_1\,\overline{z}_2, \quad \Rightarrow \quad \theta_1 = \theta_2 = \theta, \;\; w = e^{i\theta}\, z,$$

and the proof finishes. □

Let us recall what we saw in Sect. 2.2. A differentiable map between manifolds, $f : M \to N$, is said to be a *submersion* when its differential, $df_m : T_m M \to T_{f(m)} N$, is surjective at each point $m \in M$. Naturally, the dimension of M is considered greater than that of N. So the implicit function theorem guarantees that the space M admits a foliation by submanifolds with dimension $\dim(M) - \dim(N)$. Precisely, the tangent space to these submanifolds is often called the *vertical subspace*

$$\mathcal{V}_m = T_m\left(f^{-1}(f(m))\right) = \mathrm{Ker}(df_m), \qquad \text{vertical subspace.}$$

In particular, we are interested in the idea of Riemannian submersion. That is, it is considered a submersion between two Riemannian spaces

$$f : (M, \overline{g}) \to (N, g),$$

and at each point of M is taken the subspace orthogonal to the vertical subspace, which is called *horizontal subspace*

$$\mathcal{H}_m = \mathcal{V}_m^{\perp}.$$

It is obvious then that the restriction, at each point, of the differential to the horizontal subspace is an isomorphism on the tangent of N

$$df_m|_{\mathcal{H}_m} : \mathcal{H}_m \to T_{f(m)}N.$$

When it is also an *isometry* then the *submersion is said to be Riemannian*. Naturally, at each point we have the following orthogonal decomposition

$$T_m M = \mathcal{H}_m \oplus \mathcal{V}_m.$$

Usually, given a Riemannian submersion, the space M (domain) is often called the total space, while the codomain, N, is the base space. Also, in the jargon of Riemannian immersions, a widely used technique is that of horizontal liftings. Given a vector field X in the base, then a unique horizontal vector field $\overline{X}$ can be found (its representative at each point is in the horizontal subspace) in M so that

$$df_m(\overline{X}_m) = X_{f(m)}, \qquad \text{for any } m \in M.$$

It will then be said that $\overline{X}$ is *the horizontal lifting* of the field X. You can also talk about lifting of curves, simply by integration of the above vector fields. Specifically, given a curve $\alpha : [0, L) \to N$ on the base space, then for any point $m \in f^{-1}(\alpha(0))$ we have a unique curve $\overline{\alpha} : [0, L) \to M$ satisfying

(1) $\overline{\alpha}(0) = m$;
(2) The curve is horizontal, i.e., $\overline{\alpha}'(t) \in \mathcal{H}_{\overline{\alpha}(t)}$; and
(3) The curve is a horizontal lifting of the given curve, $f(\overline{\alpha}(t)) = \alpha(t)$.

We will call the curves thus obtained *horizontal liftings*.

There are two basic properties that we will use about Riemannian immersions:

(i) $[\overline{X}, V] = 0$; and
(ii) If $\overline{\nabla}$ and ∇ are the Riemannian connections, then the horizontal part of $\overline{\nabla}_{\overline{X}} \overline{Y}$ is the horizontal lifting of $\nabla_X Y$.

To learn more about Riemannian submersion, presumably the two better references are [22] and [82].

11.4 Non-massive Bosons Dynamics

The sphere $\mathbb{S}^3$ possesses a nice structural richness. First, and along with to the circle $\mathbb{S}^1$, it is the only sphere that has a group structure. Here we will exploit some aspects of its geometry, derived from the Hopf, which will be very useful to us. It is highly recommended to read H. K. Urbantke [103].

It is observed that, for each point $z \in \mathbb{S}^3$, the tangent $T_z\mathbb{S}^3$ is orthogonal to the line that generates the position vector, that is, we have the following orthogonal decomposition

$$\mathbb{C}_z = T_z\mathbb{S}^3 \oplus \{z\}, \quad \text{for any } z \in \mathbb{S}^3.$$

Moreover, we have the tangent vector fields $V(z) = i\,z$ that appeared associated to the flow of orbits by the previous action. As a consequence, we have defined a pair of distributions on $\mathbb{S}^3$, the one generated by the above field and its orthogonal. In the jargon of Riemannian submersion theory they are often called *vertical* and *horizontal* distributions, respectively.

It turns out, then, that the following orthogonal decomposition can be written

$$T_z\mathbb{S}^3 = \mathcal{H}_z \oplus \mathcal{V}_z, \qquad \mathcal{H}_z = \{i\,z\}^{\perp}, \quad \mathcal{V}_z = \{i\,z\}, \quad \text{for any } z \in \mathbb{S}^3.$$

It is clear that the horizontal distribution $\mathcal{H}_z$ is invariant because of the complex structure which is naturally defined on $\mathbb{C}^2$, that is, if $\vec{x} \in \mathcal{H}_z$, then $i\,\vec{x} \in \mathcal{H}_z$. A first conclusion is that the Hopf map is a Riemannian submersion, which means that the differential, at each point, is a surjective map and, restricted to the horizontal space, provides an isometry from this to the tangent plane to the base space

$$d\Pi_z : \mathcal{H}_z \to T_{\Pi(z)}\mathbb{S}^2.$$

Then it turns out that the Hopf map from the round 3-sphere with radius one to the round 2-sphere of radius $1/2$ is a Riemannian submersion.

An important point is that $V(z) = i\,z$ is a Killing vector field. Indeed, it suffices to realize that the one-parameter group of transformations associated to this field is

$$\{\phi_t : \mathbb{S}^3 \to \mathbb{S}^3 \ : \ \phi_t(z) = e^{it}z \quad t \in \mathbb{R}\},$$

and these transformations are isometries of the round 3-sphere. As is known, the fact that a field V on a Riemannian manifold, $(M, \langle , \rangle)$, be Killing is equivalent to its covariant derivative being antisymmetric, that is,

$$\langle \nabla_X V, Y\rangle + \langle \nabla_Y V, X\rangle = 0.$$

From now on we will represent the Riemannian connections of the spaces $\mathbb{C}^2$, $\mathbb{S}^3$ and $\mathbb{S}^2$ by D, $\overline{\nabla}$ and ∇, respectively. So, you have

$$D_X\, z = X = -A(X), \qquad X \in \chi(\mathbb{S}^3),$$

which is nothing but expressing the umbilical character of the sphere into the euclidean space.

Naturally there are equations that relate the behaviors of these connections through the horizontal liftings. The are the **O'Neill formulas**, which in the case of the of Hopf map they state as follows

$$\overline{\nabla}_{\overline{X}}\overline{Y} = \overline{\nabla_X Y} - \langle i\,\overline{X}, \overline{Y}\rangle\, V,$$
$$\overline{\nabla}_{\overline{X}} V = \overline{\nabla}_V\,\overline{X} = i\,\overline{X},$$
$$\overline{\nabla}_V\, V = 0.$$

The first equality of the second equation is clear from $[\overline{X}, V] = 0$. Furthermore, we have

$$\overline{\nabla}_{\overline{X}} V = \overline{\nabla}_{\overline{X}}(i\,z) = i\,\overline{\nabla}_{\overline{X}} z = i\,z,$$

which proves the second equation. The horizontal part of $\overline{\nabla}_{\overline{X}}\overline{Y}$ is $\overline{\nabla_X Y}$, as just noted, in any Riemannian submersion. Then, let us look at its vertical component. But

$$\langle\overline{\nabla}_{\overline{X}}\overline{Y}, V\rangle = -\langle\overline{\nabla}_{\overline{X}} V, \overline{X}\rangle = -\langle i.\overline{X}, \overline{Y}\rangle,$$

which proves the first equation. Finally, as V is a unit vector field, we have $0 = V\langle V, V\rangle = 2\langle\overline{\nabla}_V\, V, V\rangle$, so that we will have to see that the horizontal part of $\overline{\nabla}_V\, V$ vanishes. But that is obvious, since we have

$$\langle\overline{\nabla}_V\, V, \overline{X}\rangle = -\langle\overline{\nabla}_{\overline{X}} V, V\rangle = -\frac{1}{2}\,\overline{X}\langle V, V\rangle = 0.$$

It seems natural to wonder about the relationship that may exist between the Frenet apparatus of the curves in $\mathbb{S}^2$ and their horizontal liftings in $\mathbb{S}^3$. More precisely,, we consider a curve $\alpha(s)$ parameterized by the arc length in $\mathbb{S}^2$, whose Frenet apparatus is $\{T(s) = \alpha'(s), N(s)\}$ with curvature function $\kappa(s)$ and, therefore, with Frenet equations

$$\nabla_T\, T = \kappa\, N,$$
$$\nabla_T\, N = -\kappa\, T.$$

On the other hand, a horizontal lifting $\overline{\alpha}(s)$ is taken in $\mathbb{S}^3$ (which is also parameterized by the arc length, since the Hopf application is a Riemannian submersion) whose Frenet apparatus is $\{\overline{T}(s), \xi_2(s), \xi_3(s)\}$, where $\rho(s)$ and $\tau(s)$ are their curvature and torsion functions, respectively (note that their tangent field must be the horizontal lifting of $T(s)$, following the used notation). Its Frenet equations are

$$\overline{\nabla}_{\overline{T}}\overline{T} = \rho\,\xi_2,$$
$$\overline{\nabla}_{\overline{T}}\xi_2 = -\rho\,\overline{T} + \tau\,\xi_3,$$
$$\overline{\nabla}_{\overline{T}}\xi_3 = -\tau\,\xi_2.$$

Using the O'Neill equations, we can relate the corresponding geometric invariants of both curves. So from the first O'Neill equation, we get

$$\overline{\nabla}_{\overline{T}}\overline{T} = \rho\, \xi_2 = \overline{\nabla_T T} - \langle i\,\overline{T}, \overline{T}\rangle = \overline{\kappa\, N} = (\kappa \circ \Pi)\,\overline{N},$$

obtaining the following information:

(1) The curvature of a horizontal lifting is obtained by lifting the curvature of the curve in the base space, $\rho = \kappa \circ \Pi$;
(2) The principal normal of a horizontal lifting is horizontal and is obtained as the horizontal lifting of the normal of the curve in the base space, $\xi_2 = \overline{N}$;
(3) As a consequence of the above, the binormal of a horizontal lifting is tangent to the orbits.

As a consequence, we can now use again the first O'Neill equation to get

$$\overline{\nabla}_{\overline{T}}\,\xi_2 = \overline{\nabla}_{\overline{T}}\overline{N} = \overline{\nabla_T N} - \langle i\,\overline{T}, \overline{N}\rangle\, V = -(\kappa \circ \Pi)\,\overline{T} - (\langle JT, N\rangle \circ \Pi)$$
$$V = -(\kappa \circ \Pi)\,\overline{T} - V.$$

Comparing this equation with the second Frenet equation of the curve $\overline{\alpha}$ we find the following that:

(i) $\rho = \kappa \circ \Pi$, which we already know;
(ii) $\tau^2 = 1$, that is, the horizontal liftings of the curves in $\mathbb{S}^2$, are solutions of the total curvature in $\mathbb{S}^3$, in other words, they are trajectories of the non-massive bosons.

The natural question one now asks is **are there other trajectories besides horizontal liftings?**

The answer is that *the horizontal liftings form the set of all trajectories.*

Indeed, given a trajectory $\beta(s)$, we parameterize it by the arc length and consider its curvature $\rho(s)$ and torsion $\tau(s)$ functions, which should satisfiy $\tau^2(s) = 1$. Now in $\mathbb{S}^2$ we take a curve parameterized by the arc length and with curvature function $\rho(s)$ (this curve is unique up tp motions). By a horizontal lifting of this curve, we get a new curve also parameterized by the arc length whose curvature and torsion functions coincide with those of the starting curve and therefore they are congruent.

Remark In the aboove argument we have used the existence and uniqueness theorems for curves in $\mathbb{S}^2$ and $\mathbb{S}^3$, which work exactly alike as in the plane and euclidean space.

11.5 Hopf Tubes, Helices, and Massive Boson Dynamics

We will start by recalling what we saw in Sect. 2.7. Given a curve $\alpha : \mathrm{I} \subset \mathbb{R} \to \mathbb{S}^2$, we can consider its complete lifting $\Pi^{-1}(\alpha)$ in $\mathbb{S}^3$. It is clear that it is made up of all the horizontal liftings of the given curve, so that choosing one of them, $\overline{\alpha}(s)$, any other is obtained from it by the action of the group $\mathbb{S}^1$

$$\Pi^{-1}(\alpha) = \{e^{it}\,\overline{\alpha}(s)) \;:\; s \in \mathrm{I},\;\; t \in \mathbb{R}\}.$$

This provides a surface in $\mathbb{S}^3$ that can be parameterized from the following covering map

$$\mathrm{X} : \mathrm{I} \times \mathbb{R} \to \Pi^{-1}(\alpha) \subset \mathbb{S}^3, \qquad \mathrm{X}(s,t) = e^{it}\,\overline{\alpha}(s)).$$

The induced metric can be computed from that of $\mathbb{S}^3$

$$\mathrm{X}_s = e^{it}\,\overline{\alpha}'(s), \qquad \mathrm{X}_t = i\,e^{it}\,\overline{\alpha}(s),$$

to get

$$\langle \mathrm{X}_s, \mathrm{X}_s \rangle = \langle \mathrm{X}_t, \mathrm{X}_t \rangle = 1, \qquad \langle \mathrm{X}_s, \mathrm{X}_t \rangle = 0.$$

It is a flat surface (zero Gaussian curvature) which we called *Hopf tube* built on the curve α and denoted by T_α. Since X is, at least locally, an isometry, the geodesics of the Hopf tube can be viewed as images by X of segments in $\mathrm{I} \times \mathbb{R} \subset \mathbb{R}^2$. Taking one of them

$$\delta(u) = \mathrm{X}(au, bu) = e^{ibu}\,\overline{\alpha}(au),$$

we can calculate the Frenet apparatus of this curve in $\mathbb{S}^3$

$$\delta'(u) = ae^{ibu}\,\overline{\alpha}'(au) + bie^{ibu}\,\overline{\alpha}(au) = ae^{ibu}\,\overline{\alpha}'(au) + b\,V(\delta(u)),$$

so $\langle \delta'(u), \delta'(u) \rangle = a^2 + b^2$ and therefore its arc length parameter satisfies $\dfrac{d\overline{s}}{du} = \sqrt{a^2+b^2}$ so that its unit tangent is

$$T = \frac{1}{\sqrt{a^2+b^2}}\,\delta' = \frac{a}{\sqrt{a^2+b^2}}\,e^{ibu}\,\overline{\alpha}' + \frac{b}{\sqrt{a^2+b^2}}\,V.$$

From here, using O'Neill formulas, we can compute the curvature and principal normal, in $\mathbb{S}^3$, of the geodesic of the Hopf tube:

$$\overline{\nabla}_T\,T = \frac{a^2}{a^2+b^2}\,\overline{\nabla_{\alpha'}\,\alpha'} + \frac{2ab}{a^2+b^2}\,i\,e^{ibu}\,\overline{\alpha}'.$$

Now remembering that $i e^{ibu}\,\overline{\alpha}'$ is the horizontal lifting of the normal of the curve α, we have

$$\overline{\nabla}_T T = \frac{a^2\,\kappa \circ \Pi + 2ab}{a^2+b^2}\, i e^{ibu}\,\overline{\alpha}',$$

and therefore the curvature, in $\mathbb{S}^3$, of the geodesic δ of the Hopf tube T_α and the curvature of α in $\mathbb{S}^2$ are related as follows

$$\rho^2 = \left(\frac{a^2\,\kappa \circ \Pi + 2ab}{a^2+b^2}\right)^2.$$

Thus, a first important consequence is obtained:

Geodesics of Hopf tubes on circles of $\mathbb{S}^2$ have constant curvature in $\mathbb{S}^3$.

The question that now naturally appears to us is that of considering Hopf tubes on circles of $\mathbb{S}^2$ and see if their geodesics also have constant torsion. In such case, they should be helices in $\mathbb{S}^3$. Furthermore, it must be observed that the constants a and b define the slope $\omega = b/a$ of the straigh line chosen such that

$$\rho^2 = \left(\frac{\kappa \circ \Pi + 2\omega}{1+\omega^2}\right)^2.$$

As we have just seen

$$T = \frac{a}{\sqrt{a^2+b^2}}\, e^{ibu}\,\overline{\alpha}' + \frac{b}{\sqrt{a^2+b^2}}\, V, \qquad N = i\, e^{ibu}\,\overline{\alpha}'.$$

Thus, if we consider the orientation in $\mathbb{S}^3$ so that

$$(e^{ibu}\,\overline{\alpha}') \times (i\, e^{ibu}\overline{\alpha}') = V,$$

we have that the unit binormal is

$$B = T \times N = \frac{a}{\sqrt{a^2+b^2}}\, V - \frac{b}{\sqrt{a^2+b^2}}\, e^{ibu}\,\overline{\alpha}'.$$

Now we take derivative and apply the third Frenet equation to calculate the torsion as

$$\overline{\nabla}_T B = -\tau\, N = \frac{a^2-b^2}{a^2+b^2}\, i\, e^{ibu}\,\overline{\alpha}' - \frac{ab}{a^2+b^2}\, \overline{\nabla_T T},$$

$$-\,\tau\, N = \left(\frac{a^2-b^2}{a^2+b^2} - \frac{ab(\kappa \circ \Pi)}{a^2+b^2}\right) i\, e^{ibu}\,\overline{\alpha}'.$$

Therefore we have

$$\tau^2 = \left(\frac{a^2 - b^2}{a^2 + b^2} - \frac{ab(\kappa \circ \Pi)}{a^2 + b^2}\right)^2 .$$

As a consequence, we have obtained the following result

Geodesics of Hopf tubes on circles of $\mathbb{S}^2$ are helices in $\mathbb{S}^3$.

Now, it seems reasonable to ask ourselves **are there other helices in $\mathbb{S}^3$?**

Let us see that the answer is negative. To do that we first see that in the same way that the curvature and torsion of every geodesic of a Hopf tube on a circle, are determined from the curvature of the circle and the slope of the geodesic, these two parameters can be determined from the curvature and torsion.

In other words, we will prove the following result:

Any geodesic of a Hopf tube on a circle is determined in a unique way, up to motions of $\mathbb{S}^3$, by one of the following two pairs of parameters

- **Its curvature, ρ, and its torsion, τ;**
- **The curvature κ of the circle in $\mathbb{S}^2$ and its slope ω in the Hopf tube.**

To prove that we start from the expressions that determine (ρ, τ) as a function of (κ, ω)

$$\rho = \frac{\kappa + 2\omega}{1 + \omega^2}, \qquad \tau = \frac{1 - \omega^2 - \omega\kappa}{1 + \omega^2}.$$

Let us see the converse. From the first equation we have

$$\kappa = (1 + \omega^2)\rho - 2\omega,$$

which we bring to the second one to obtain

$$\omega = \frac{1 - \tau}{\rho},$$

that substituted in the previous one provides the following value of κ in terms of (ρ, τ)

$$\kappa = \frac{\rho^2 + \tau^2 - 1}{\rho}.$$

Naturally, all of the above along with the fundamental theorem of curves on $\mathbb{S}^3$ gives the result.

A pair of final remarks:

(1) **As we have just known, the dynamics associated with the massive bosons in $\mathbb{S}^3$ is governed by the Lagrangian**

$$\mathcal{F}_m(\gamma) = \int_\gamma (\rho + m)\, ds, \qquad m \in \mathbb{R},$$

so that the trajectories are the helices with curvature ρ and torsion τ, satisfying

$$\tau^2 + m\,\rho = 1.$$

This equation can be written in terms of the parameters κ (curvature in $\mathbb{S}^2$ of the circle on which the Hopf tube is constructed) and ω (slope of the helix in the Hopf tube).

(2) **In the same way that the trajectories of the model can be described in terms of either (ρ, τ) or (κ, ω), there is another way to describe the trajectories in terms of other two parameters, the mass and spin (see [45] and [65]).**

Appendix A
The Standard 3-Sphere $\mathbb{S}^3$

The Hopf map has allowed us an intense contact with $\mathbb{S}^3$, but due to the key importance of this space, much beyond the content of this work, we present a beautiful study of $\mathbb{S}^3$ due to Louis Zulli [110] which will lead us to understand the analogies and differences between this rich sphere and the more known $\mathbb{S}^2$.

Let $q = x+yi+zj+tk$ be a quaternion which, when convenient, we will identify with the point (x, y, z, t) of $\mathbb{R}^4$ and, sometimes, with the vector $\vec{q} = (v_1, v_2) \in \mathbb{C}^2$, where $v_1 = x + yi$ and $v_2 = z + ti$. There is a correspondence between the quaternions and certain 2×2 matrices of complex numbers of the form

$$q = x + yi + zj + tk \in \mathbf{H} \equiv \mathbb{R}^4 \longleftrightarrow V = \begin{pmatrix} v_1 & v_2 \\ -\overline{v}_2 & \overline{v}_1 \end{pmatrix} \in M_2(\mathbb{C}).$$

This correspondence, which preserves the sum and the product, is a ring division isomorphism that allows us to obtain a matrix model of $\mathbb{R}^4$. It is immediate to verify that, in this model, the euclidean scalar product $\langle V_1, V_2\rangle = x_1x_2+y_1y_2+z_1z_2+t_1t_2$ is written as $\langle V_1, V_2\rangle = (1/2)\mathrm{tr}(V_1 \bar{V}_2^T)$ and $|V|^2 = \det(V)$.

By definition, $\mathbb{S}^3 = \{q \in \mathbf{H} : |q| = 1\} = \{\vec{q} \in \mathbb{R}^4 : |\vec{q}| = 1\}$. We know that $\mathbb{S}^3$ is a non-commutative group via the quaternions multiplication. In the matrix model of $\mathbb{R}^4$, we have the following map

$$\begin{aligned} \mathbb{S}^3 &\hookrightarrow M_2(\mathbb{C}) \\ V &\mapsto \begin{pmatrix} v_1 & v_2 \\ -\overline{v}_2 & \overline{v}_1 \end{pmatrix} : \det(V) = 1. \end{aligned}$$

The image set can also be described by $\{M \in M_2(\mathbb{C}) : M\overline{M}^T = I,\ \det(M) = 1\}$. This matrix group is represented by $SU(2)$ and is called the 2-dimensional special unitary group. The correspondence $\mathbb{S}^3 \leftrightarrow SU(2)$ is a group isomorphism. Therefore, sometimes it is not easy to distinguish between $\mathbb{S}^3$ and $SU(2)$. In

A. Ferrandez Izquierdo, *Helices*, RSME Springer Series 16,
https://doi.org/10.1007/978-3-032-09663-0

particular, the points of $\mathbb{S}^3$, seen as matrices, will be represented by uppercase letters.

A.1 Distance, Parallels and Meridians in $\mathbb{S}^3$

Distance The distance in $\mathbb{R}^4$ between two points $v, w \in \mathbb{S}^3$ is the usual one, that is, $|v - w|$, which in matrix formulation is written as $\sqrt{\det(V - W)}$. This distance is, simply, the length of a vector in $\mathbb{R}^4$ from v to w. But such a vector intersects $\mathbb{S}^3$ only in v and w and its length is not the distance in $\mathbb{S}^3$ between v and w.

To compute $d_{\mathbb{S}^3}(v, w)$, the intrinsic distance from v to w in $\mathbb{S}^3$, proceed as follows. Take a 2-plane Π that, passing through the origin O of $\mathbb{R}^4$, contains v and w. That plane will be unique, up to $w = -v$, but uniqueness is not required. Let us see what is $\Pi \cap \mathbb{S}^3$ in $\mathbb{R}^4$. The intersection $\Pi \cap \mathbb{S}^3$ is in both Π and $\mathbb{S}^3$, but it is easier to see it into Π, where it is exactly the set of points of Π at a distance 1 from the origin $O \in \mathbb{R}^4$, that is, $\Pi \cap \mathbb{S}^3$ is the unit circle of the plane Π. In $\mathbb{S}^3$, the circle $\Pi \cap \mathbb{S}^3$ is a great circle and the key fact we need is that it performs the shortest distance in $\mathbb{S}^3$ from v to w. Then it is easy to compute $d_{\mathbb{S}^3}(v, w)$ by setting our attention in the plane Π. Since v and w are on the unit circle in Π, the distance between them on the circle is given by the (smallest) angle $\theta = \sphericalangle(vOw)$, measured in radians, in the plane Π (see Fig. A.1).

Since v and w are unitary, we have the classical formula $\theta = \arccos(\langle v, w\rangle)$. So, $d_{\mathbb{S}^3}(v, w) = \arccos(\langle v, w\rangle)$, or in matrix form

$$d_{\mathbb{S}^3}(v, w) = \arccos\big(\frac{1}{2}\mathrm{tr}(V\overline{W}^T)\big).$$

Parallels Let us keep this formula for the distance in $\mathbb{S}^3$. Since the identity matrix I corresponds to the point $(1, 0, 0, 0) \in \mathbb{S}^3 \subset \mathbb{R}^4$, the x axis of $\mathbb{R}^4$ passes through I. For convenience, we will see the point I as the "North Pole" of our model of $\mathbb{S}^3$ and $-I$ as the "South Pole". For $L \in [0, \pi]$, let $\mathcal{P}_L = \{V \in \mathbb{S}^3 : d_{\mathbb{S}^3}(V, I) = L\}$. We will call $\mathcal{P}_L$ the **parallel in $\mathbb{S}^3$ of latitude** L. Note that referring to distance L as latitude is quite reasonable, since it is not only L the distance in $\mathbb{S}^3$ from V to I, but also is the radian measure of angle $\sphericalangle(IOV)$. Observe that we are measuring latitude starting from the North Pole, that is to say, contrary to what is usual on Earth, where it is measured from the Equator. So, $\mathcal{P}_0 = \{I\}$, i.e., I is the only point of $\mathbb{S}^3$ of latitude 0, since it is the unique point of $\mathbb{S}^3$ whose distance to I is 0. Also,

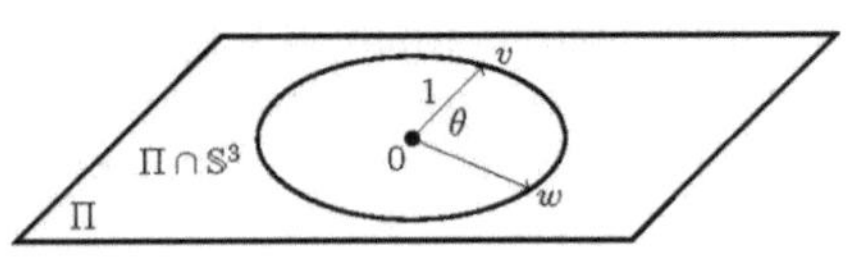

Fig. A.1 Distance on the unit circle

$\mathcal{P}_\pi = \{-I\}$, and we will refer to $\mathcal{P}_{\pi/2}$ as the Equator of $\mathbb{S}^3$, which being a special parallel it will be denoted by $\mathcal{E}$.

It will be very useful to have on hand different descriptions of $\mathcal{P}_L$. It is immediately checked the following

$$\mathcal{P}_L = \{V \in \mathbb{S}^3 \,:\, (1/2)\mathrm{tr}(V) = \cos L\},$$

$$\mathcal{P}_L = \{(x, y, z, t) \in \mathbb{S}^3 \subset \mathbb{R}^4 \,:\, x = \cos L\}.$$

Indeed,

$$V\overline{I}^T = \begin{pmatrix} v_1 & v_2 \\ -\overline{v}_2 & \overline{v}_1 \end{pmatrix} \begin{pmatrix} 1 & 0 \\ 0 & 1 \end{pmatrix} = V.$$

Therefore, $\mathrm{tr}(V\overline{I}^T) = \mathrm{tr}(V) = v_1 + \overline{v}_1 = 2x$, where $v_1 = x + yi$ and $v_2 = z + ti$.

From these expressions the following consequences can be deduced:

1. From the first expression of $\mathcal{P}_L$ it can be concluded that all the points of a given parallel have the same trace and that *V is on the equator $\mathcal{E}$ of $\mathbb{S}^3$ if and only if* $\mathrm{tr}(V) = 0$. That means that if we write $V = \begin{pmatrix} w & * \\ * & \overline{w} \end{pmatrix}$, we have that $\mathrm{tr}(V) = 2\mathrm{Re}(w)$. As $\mathrm{tr}(V) = 0$, it should be that $\mathrm{Re}(w) = 0$.
2. From the second formulation, we see that $\mathcal{P}_L$ is a 2-sphere of radius $\sin L$ centered at the point $(\cos L, 0, 0, 0)$ in the hyperplane $x = \cos L$ in $\mathbb{R}^4$. (If $L = 0$ or $L = \pi$ that 2-sphere has radius 0, and so $\mathcal{P}_0$ and $\mathcal{P}_\pi$ are single point sets). In particular, $\mathcal{E}$ is the 2-unit sphere in the 3-dimensional subspace of $\mathbb{R}^4$ spanned by $\{y, z, t\}$, or equivalently, $\mathcal{E}$ is the set of unit pure imaginary quaternions, that is, those of the form $yi + zj + tk$.
3. Also from the second formulation of $\mathcal{P}_L$ it can be concluded that $\mathbb{S}^3 = \bigcup_{L\in[0,\pi]} \mathcal{P}_L$, that is, each point of $\mathbb{S}^3$ is in some parallel. This is because $x \in [-1, 1]$ for each point $(x, y, z, t) \in \mathbb{S}^3 \subset \mathbb{R}^4$, and so $x = \cos L$ for some $L \in [0, \pi]$. It is clear that $\mathcal{P}_L \cap \mathcal{P}_{L'} = \emptyset$ when $L \neq L'$, and so each point of $\mathbb{S}^3$ lies in a unique parallel.

Meridians Let us now take a point $E \in \mathcal{E}$. By the **meridian $\mathbb{M}_E$ of longitude** E we will understand the set

$$\mathbb{M}_E = \{V \in \mathbb{S}^3 \,:\, V = \cos(\theta)I + \sin(\theta)E,\ \theta \in [0, \pi]\}.$$

Note that, as in the case of the parallels of $\mathbb{S}^3$, we do not follow the measurement convention of the terrestrial length, because for us the length of a meridian in $\mathbb{S}^3$ is simply the intersection point of such a meridian with the Equator. If this idea were transferred to Earth, then the meridian that passes through Buffalo would be called "Quito, Ecuador" instead of "$78^o W$".

It is easy to check:

(1) $\mathbb{M}_E$ and $\mathcal{P}_L$ intersect each other at the point $\cos(L)I + \sin(L)E$.
Indeed, $\mathbb{M}_L$ is the semicircle $\cos(\theta)I + \sin(\theta)E$ from the North Pole to the South Pole, which will intersect the parallel $\mathcal{P}_L$ when $\theta = L$.

(2) $\mathbb{M}_E \cap \mathbb{M}_{E'} = \{\pm I\}$, when $E \neq E'$.
Indeed, if they were intersected at another point W, then $W \in \mathcal{P}_{L_W}$, so that $\mathbb{M}_E \cap \mathcal{P}_{L_W} = W$ and $\mathbb{M}_{E'} \cap \mathcal{P}_{L_W} = W$, that is, $\cos(L_W)I + \sin(L_W)E = \cos(L_W)I + \sin(L_W)E'$, which cannot hold because $E \neq E'$.

To see that $\mathbb{S}^3 = \bigcup_{E\in\mathcal{E}} \mathbb{M}_E$, that is, that every point $V \in \mathbb{S}^3$ is in a meridian, we will do the following geometric argument. Let us choose a plane Π in $\mathbb{R}^4$ containing I, V and the origin O. As before, Π will be unique if $V \neq \pm I$, but the uniqueness of Π is not needed. Let $\mathbb{R}^3$ the 3-dimensional subspace of $\mathbb{R}^4$ spanned by $\{y, z, t\}$, so that $\mathbb{R}^3 \cap \mathbb{S}^3 = \mathcal{E}$. It is clear that Π is not a subspace of $\mathbb{R}^3$, since $I \in \Pi$ but $I \notin \mathbb{R}^3$. As $\Pi \cap \mathbb{R}^3$ is necessarily a subspace of Π, then either $\Pi \cap \mathbb{R}^3 = \{O\}$, or $\Pi\cap\mathbb{R}^3$ is a line through the origin O. The first possibility is eliminated by dimension considerations, therefore only the second one remains. Such straight line cuts $\mathcal{E}$, the unit 2-sphere in $\mathbb{R}^3$, at two points, E and $-E$. As $(\Pi\cap\mathbb{R}^3)\cap\mathcal{E} = (\Pi\cap\mathbb{S}^3)\cap\mathcal{E}$, the great circle $\Pi \cap \mathbb{S}^3$, which contains V, cuts $\mathcal{E}$ at $\pm E$. Since I and E are orthogonal points on this circle in Π, each point on the circle can be expressed in the form $\cos(\theta)I + \sin(\theta)E$, for some $\theta \in [0, 2\pi]$. In particular, $V = \cos(\theta)I + \sin(\theta)E$, for some $\theta \in [0, 2\pi]$. Then $\theta \in [0, \pi]$ and $V \in \mathbb{M}_E$, and we finish. On the contrary, $\theta = 2\pi - \theta'$, with $\theta' \in [0, \pi]$, and substituting we get $V = \cos(\theta')I + \sin(\theta')(-E)$, so $V \in \mathbb{M}_{-E}$.

Although this geometric argument proves that each $V \in \mathbb{S}^3$ is in a meridian, however it does not provide the best procedure for calculate the point E (or $-E$) that fixes the meridian. In what follows, we will see another way to find the meridian in question. For now, we have established that each point $V \in \mathbb{S}^3 - \{\pm I\}$ lies in a unique parallel $\mathcal{P}_L$ and a unique meridian $\mathbb{M}_E$, and that $\mathcal{P}_L \cap \mathbb{M}_E = \{V\}$. Thus, each V has a unique latitude $L \in [0, \pi]$ and a unique longitude $E \in \mathcal{E}$, and also, no other point of $\mathbb{S}^3$ have these latitude and longitude.

A.2 Conjugation in $\mathbb{S}^3$

The sphere $\mathbb{S}^3$ is much more interesting than $\mathbb{S}^1$ or $\mathbb{S}^2$, because their structure group, as we will see, fits very well with its geometry. (We know that $\mathbb{S}^1$ is also a group, but its multiplication is commutative, which makes that its overall structure is less interesting than that of $\mathbb{S}^3$. And $\mathbb{S}^2$ admits no group structure analogous to that of the other spheres). In particular, we want to understand how the conjugation in $\mathbb{S}^3$

interacts with the coordinate systems we built before. To do this, given $C \in \mathbb{S}^3$, the application is defined

$$\begin{aligned} C_* : \mathbb{S}^3 &\to \mathbb{S}^3 \\ V &\mapsto CVC^{-1}, \end{aligned}$$

which we will call *conjugation* for C.

Since $C \in \mathbb{S}^3$, we have that $C\bar{C}^T = I$, it turns out that also you can write

$$C_*(V) = CV\bar{C}^T.$$

It is easy to see that, for every $C \in \mathbb{S}^3$, C_* is a homomorphism and that $(C_*)^{-1} = (C^{-1})_*$, therefore C_* is an isomorphism of groups. But even better, because C_* is an isometry of $\mathbb{S}^3$, that is, for any $V, W \in \mathbb{S}^3$

$$d_{\mathbb{S}^3}(C_*(V), C_*(W)) = d_{\mathbb{S}^3}(V, W).$$

Indeed, a long and tedious calculation allows us to verify that the conjugation preserves the trace of a matrix.

Since it is an isometry and $\mathcal{P}_L = \{V \in \mathbb{S}^3 : d_{\mathbb{S}^3}(I, V) = L\}$, C_* maps $\mathcal{P}_L$ isometrically to itself for each $L \in [0, \pi]$ and each $C \in \mathbb{S}^3$. In fact, the collection of parallels of $\mathbb{S}^3$ is precisely the set of conjugation classes in $\mathbb{S}^3$, that is, *V and W are conjugate in* $\mathbb{S}^3$ (meaning that $W = C_*(V)$, *for some* $C \in \mathbb{S}^3$) *if, and only if, V and W are in the same parallel, that is, if and only if V and W they have the same latitude at* $\mathbb{S}^3$. Before to show this, it will be useful to extend the terminology and notation. For $L \in [0, \pi]$, we take the point $D_L = \begin{pmatrix} e^{iL} & 0 \\ 0 & e^{-iL} \end{pmatrix} \in \mathbb{S}^3$, and the set $\mathcal{D} = \{D_L : L \in [0, \pi]\}$. As $D_{\pi/2} = \begin{pmatrix} i & 0 \\ 0 & -i \end{pmatrix}$, it turns out that $D_{\pi/2}$ is a point on the equator $\mathcal{E}$ and then $\mathcal{D}$ coincides with the meridian $\mathbb{M}_{D_{\pi/2}}$. We will refer to $\mathcal{D}$ as the *principal meridian* of $\mathbb{S}^3$. (We use the letter $\mathcal{D}$ because all points D_L of $\mathcal{D}$ are diagonal matrices). Since $D_{\pi/2}$ corresponds to the point $(0, 1, 0, 0)$ of $\mathbb{R}^4$, the y axis passes through such a point and thus $\mathcal{D}$ is contained in the unit circle $\mathbb{S}^1 \subset \mathbb{R}^2$. In fact, $\mathcal{D}$ is precisely the unit semicircle in the closed upper half-plane of $\mathbb{R}^2$. It is worth noting that for each $L \in [0, \pi]$, $\mathcal{D}$ intersects the parallel $\mathcal{P}_L$ at the point D_L. In particular, the latitude of D_L is L.

Let us see that $\mathcal{P}_L$ is precisely the conjugation class of D_L in $\mathbb{S}^3$. We already know that every conjugate of D_L is in $\mathcal{P}_L$. What we need to prove is that each point V of $\mathcal{P}_L$ is a conjugate of the diagonal matrix D_L, that is, $C_*(D_L) = V$. This is done as follows: first, it is clear that the result is true for $L = 0$ and $L = \pi$, so we will assume that $L \in (0, \pi)$. Since V is a 2×2 matrix, its characteristic polynomial is $\lambda^2 - \mathrm{tr}(V)\lambda + \det(V)$. Since $V \in \mathcal{P}_L \subset \mathbb{S}^3$, then it can be written in the form $\begin{pmatrix} \cos L + yi & z + ti \\ -z + ti & \cos L - yi \end{pmatrix}$ and the polynomial becomes $\lambda^2 - (2\cos(L))\lambda + 1$, so the two distinct eigenvalues are $\lambda_1 = \cos(L) + i\sin(L) = e^{iL}$ and $\lambda_2 = \bar{\lambda}_1 = \cos(L) - i\sin(L) = e^{-iL}$. Let us equip the complex plane $\mathbb{C}^2$ with its usual Hermitian metric and let us choose a unit eigenvector (z_1, z_2) of V corresponding to the eigenvalue λ_1. Set $C = \begin{pmatrix} z_1 & -\bar{z}_2 \\ z_2 & \bar{z}_1 \end{pmatrix}$. Then $\det(C) = 1$, because (z_1, z_2) is a unit vector in $\mathbb{C}^2$, and so $C \in \mathbb{S}^3$. As

$$(C^{-1}VC)\begin{pmatrix} 1 \\ 0 \end{pmatrix} = (C^{-1}V)\begin{pmatrix} z_1 \\ z_2 \end{pmatrix} = (\lambda_1 C^{-1})\begin{pmatrix} z_1 \\ z_2 \end{pmatrix} = \lambda_1\begin{pmatrix} 1 \\ 0 \end{pmatrix} = \begin{pmatrix} \lambda_1 \\ 0 \end{pmatrix},$$

it turns out that $C^{-1}VC = \begin{pmatrix} \lambda_1 & \alpha \\ 0 & \beta \end{pmatrix}$, for some $\alpha, \beta \in \mathbb{C}$. But $C^{-1}VC \in \mathbb{S}^3$, so it must be $\alpha = 0$ and $\beta = \bar{\lambda}_1$. That is, $C^{-1}VC = D_L$, or equivalently, $V = CD_LC^{-1}$, which was were we asking for. (It should be noted that the point $C \in \mathbb{S}^3$ that conjugates D_L to V is not unique. In fact, the set of points in $\mathbb{S}^3$ that conjugate D_L to V is precisely $C\mathbb{S}^1$, the class that contains C of the subgroup $\mathbb{S}^1$ of $\mathbb{S}^3$).

After analyzing the effect of conjugation on the parallels of $\mathbb{S}^3$, let us turn our attention back to the meridians. Here there is less work.

The key result is the following: *for any $C \in \mathbb{S}^3$ and $E \in \mathcal{E}$,*

$$C_*(\mathbb{M}_E) = \mathbb{M}_{C_*(E)}.$$

In other words, the conjugation for C takes the meridian of longitude E isometrically to the meridian of longitude $C_*(E)$.

Indeed, let $V = \cos(\theta)I + \sin(\theta)E \in \mathbb{M}_E$. Then

$$C_*(V) = \cos(\theta)C_*(I) + \sin(\theta)C_*(E) = \cos(\theta)I + \sin(\theta)C_*(E),$$

i.e., $C_*(E) \in \mathbb{M}_{C_*(E)}$.

Thus, for each $C \in \mathbb{S}^3$, C_* can be thought of as a "rigid rotation" of $\mathbb{S}^3$ around its polar axis (see the Fig. A.2).

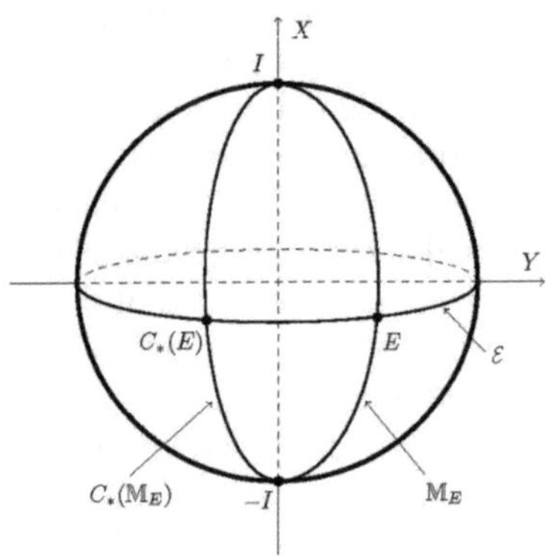

Fig. A.2 A rigid rotation of $\mathbb{S}^3$

Putting all of the above together, we have shown that, for each $C \in \mathbb{S}^3$, the conjugation for C takes the point of latitude-longitude (L, E) to the point of latitude-longitude $(L, C_*(E))$.

Using the conjugation, a second approximation can be outlined to find the length of an arbitrary point $V \in \mathbb{S}^3 - \{\pm I\}$. Let L be the latitude of V. As proved before, there exists a $C \in \mathbb{S}^3$ for which $C_*(D_L) = V$. Let $E = C_*(D_{\pi/2})$. So E has the same length as V, since D_L has the same length as $D_{\pi/2}$ and the conjugation takes meridians to meridians. Also, $E \in \mathcal{E}$, since $D_{\pi/2} \in \mathcal{E}$ and the conjugation preserves latitudes. Therefore, we have an expression of E of the form

$$\begin{aligned} E = C_*(D_L) &= \begin{pmatrix} z_1 & z_2 \\ -\overline{z}_2 & \overline{z}_1 \end{pmatrix} \begin{pmatrix} i & 0 \\ 0 & -i \end{pmatrix} \begin{pmatrix} \overline{z}_1 & -z_2 \\ \overline{z}_2 & z_1 \end{pmatrix} \\ &= i \begin{pmatrix} z_1\overline{z}_1 - z_2\overline{z}_2 & 2z_1\overline{z}_2 \\ 2\overline{z}_1 z_2 & -(z_1\overline{z}_1 - z_2\overline{z}_2) \end{pmatrix}. \end{aligned}$$

Appendix B
An Overall View of the Hopf Map

B.1 The Hopf Map, or Hopf Fibration

We will dedicate this section to the study and applications of the Hopf fibration. This map, both for its importance in itself and for its applications, should form part of the cultural heritage of our readers and contribute to our study becoming more self-contained.

The Hopf map, or fibration, so known from the 1931 Heinz Hopf work [57], is a very important tool in mathematics and physics. It was a landmark discovery for both topology and Lie group theory, and has a wide range of physical applications, from magnetic monopoles to rigid-body mechanics to quantum information theory (see [74, 76, 77] and [103]).

There are many ways to define this map, but here we have chosen the one by Lyons [72].

Definition B.1.1 (First Definition) The Hopf map $h : \mathbb{S}^3 \to \mathbb{S}^2$ is defined by

$$h(a, b, c, d) = (a^2 + b^2 - c^2 - d^2,\ 2(ad + bc),\ 2(bd - ac)). \tag{B.1}$$

It is usual to see $\mathbb{S}^3$ within $\mathbb{R}^4$, but now we will place it within the set $\mathbf{H}$ of quaternions that, as we know, both as a set as well as a vector space, it is identical to $\mathbb{R}^4$. Remember that in $\mathbf{H}$ there are three distinguished vectors $(0, 1, 0, 0)$, $(0, 0, 1, 0)$ and $(0, 0, 0, 1)$ which are often called i, j and k, respectively. The vector (a, b, c, d) is written in the form $a + bi + cj + dk$ when viewed as a quaternion. The number a is said to be the *real part* and b, c, and d are the parts i, j, and k, respectively. As in the real and complex cases, quaternions can be multiplied, but using the appropriate rules, namely,

$$i^2 = j^2 = k^2 = -1,$$

A. Ferrandez Izquierdo, *Helices*, RSME Springer Series 16,
https://doi.org/10.1007/978-3-032-09663-0

Fig. B.1 A rotation in $\mathbb{R}^3$

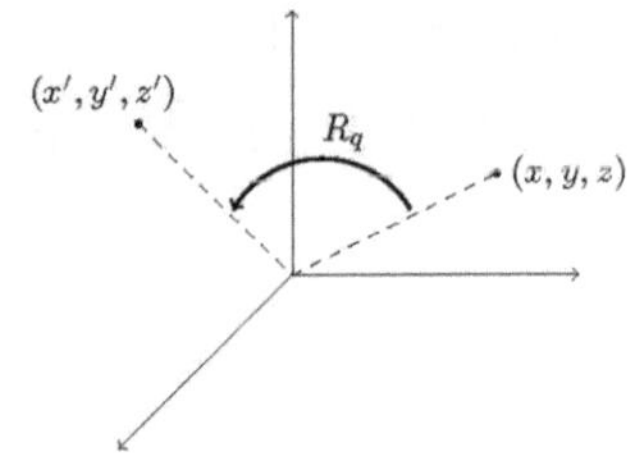

$$ij = k,\ jk = i,\ ki = j,$$

and

$$ji = -k,\ kj = -i,\ ik = -j.$$

Note that multiplication is associative, but not commutative.

We also define the *conjugate* $\overline{q}$ of a quaternion $q = a + bi + cj + dk$ by $\overline{q} = a - bi - cj - dk$ and the length, or norm, $||q||$ as its length as a vector of $\mathbb{R}^4$, that is, $||q|| = \sqrt{a^2 + b^2 + c^2 + d^2}$. We can also write $||q|| = \sqrt{q\,\overline{q}}$. Furthermore, it happens that $||q_1 q_2|| = ||q_1||\,||q_2||$. Therefore, the product of two unit quaternions (i.e., of length equal to 1) is a unit quaternion. The set of unit quaternions, as points of $\mathbb{R}^4$, is the sphere $\mathbb{S}^3$. Each nonzero quaternion q has a *multiplicative inverse* q^{-1} given by $q^{-1} = \dfrac{\overline{q}}{||q||^2}$.

Each quaternion q determines a *rotation*, that is, a linear map $R_q : \mathbb{R}^3 \to \mathbb{R}^3$ defined as follows. A point $p = (x, y, z) \in \mathbb{R}^3$ is associated to the quaternion $p = xi + yj + zk$, called *pure*, since it has no real part. The quaternion product qpq^{-1} is also pure, that is, of the form $qpq^{-1} = x'i + y'j + z'k$, and can therefore be seen as a point (x', y', z') of $\mathbb{R}^3$. Then the map R_q is defined by (see Fig. B.1)

$$R_q(x, y, z) = (x', y', z').$$

In matrix form R_q is written as

$$\begin{pmatrix} x' \\ y' \\ z' \end{pmatrix} = \begin{pmatrix} a^2 + b^2 - c^2 - d^2 & 2(bc - ad) & 2(ac + bd) \\ 2(ad + bc) & a^2 - b^2 + c^2 - d^2 & 2(cd - ab) \\ 2(bd - ac) & 2(ab + cd) & a^2 - b^2 - c^2 + d^2 \end{pmatrix} \begin{pmatrix} x \\ y \\ z \end{pmatrix}.$$

The properties of the rotations are immediate: (i) For any $a \in \mathbb{R},\ a \neq 0$, $R_{aq} = R_q$; (ii) $R_{q^{-1}} = (R_q)^{-1}$; and (iii) $R_{q_1} \circ R_{q_2} = R_{q_1 q_2}$.

Therefore, whenever $q \neq 0$, we can choose unitary q to define the rotation R_q. Furthermore, if q is a unit quaternion, both the axis and the angle of rotation are encoded in the four coordinates (a, b, c, d) as follows. If $q = 1$, then R_q is the identity map of $\mathbb{R}^3$. In another case, R_q is a rotation around the axis determined by the vector (b, c, d), with angle of rotation $\theta = 2\arccos(a) = 2\arcsin(\sqrt{b^2 + c^2 + d^2})$.

Indeed, first we have that $||R_q(p)|| = ||p||$, for any pure quaternion $p = xi + yj + zk$. Secondly, the linear map R_q has (b, c, d) as eigenvector with eigenvalue 1. Finally, to calculate the rotation angle, let us take a vector $\vec{w}$ perpendicular to the vector (b, c, d), which can be done in two ways. If at least one of b, c does not vanish, we choose $\vec{w} = ci - dj$. If $b = c = 0$, we take $\vec{w} = i$. Now the rotation angle between the vectors $\vec{w}$ and $R_q\vec{w}$ is calculated with the usual formula

$$\cos\theta = \frac{\langle \vec{w}, R_q\vec{w} \rangle}{||\vec{w}||^2} = a^2 - b^2 - c^2 - d^2 = 2a^2 - 1,$$

i.e., $a = \cos(\theta/2)$.

Remark B.1.2 $\mathbb{S}^3$, with the quaternion multiplication, has a group structure. The set of rotations in $\mathbb{R}^3$, with the composition operation, is also a group, represented by SO = (3). The map $\varphi : \mathbb{S}^3 \to SO(3)$, defined by $q \mapsto R_q$, is a group homophism. Each rotation $R \in SO(3)$ can be written in the form $R = R_q$, for some $q \in \mathbb{S}^3$, that is, φ is surjective. And each rotation R_q has two preimages in $\mathbb{S}^3$, namely q and $-q$. The kernel of φ is the subgroup $\{1, -1\}$, so we have the group isomorphism $\mathbb{S}^3/\{1, -1\} \approx SO(3)$.

B.2 Second Definition of the Hopf Map

We are going to take advantage of all we know about quaternions to reformulate the Hopf map. First, let us fix a distinguished point of $\mathbb{S}^2$, say $P_0 = (1, 0, 0)$. Given any point $(a, b, c, d) \in \mathbb{S}^3$, let $q = a + bi + cj + dk$ be the corresponding quaternion, which -as we have just seen- we know it defines an R_q rotation of $\mathbb{R}^3$. We then give a new definition of the Hopf application.

Definition B.2.1 (Second Definition) The Hopf map $h : \mathbb{S}^3 \to \mathbb{S}^2$ is defined by (see Fig. B.2)

$$q \mapsto R_q(P_0) = P = q\, i\, \overline{q}. \tag{B.2}$$

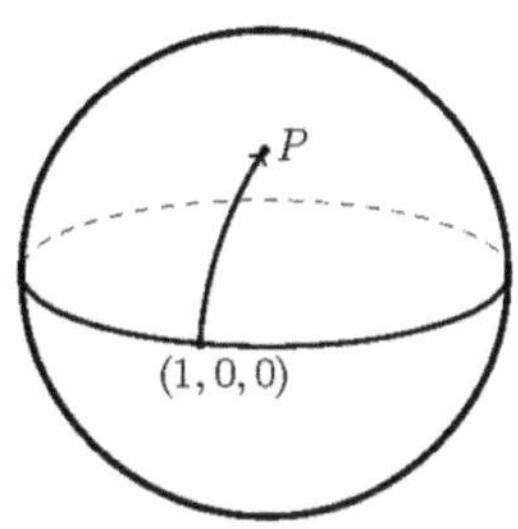

Fig. B.2 A rotation R_q

It is straightforward to check that (B.1) and (B.2) are equivalents. Indeed, applying the matrix expression of R_q we have that $h(q) = R_q((1,0,0)) = (a^2 + b^2 - c^2 - d^2,\ 2(ad + bc),\ 2(bd - ac))$.

Let us see how the fibers of this application are, that is, how are the sets of the form $h^{-1}(P)$.

If we take $P_0 = (1, 0, 0)$, it is immediate to check that

$$h^{-1}(P_0) = \{(\cos t, \sin t, 0, 0)\ :\ t \in \mathbb{R}\} \subset \mathbb{S}^3,$$

which is $\mathbb{S}^1$ lying in a plane of $\mathbb{R}^4$. In general, for any $P \in \mathbb{S}^2$, $h^{-1}(P)$, called *the fiber* of h over P, is a circle in $\mathbb{S}^3$

B.3 Configuration of the Fibres of the Hopf Map

We will try to see how the fibers of h are organized in $\mathbb{S}^3$, that is, how is the geometry of the Hopf fibration.

We have seen that h takes a point $q \in \mathbb{S}^3$ to a point $P \in \mathbb{S}^2$ through the rotation R_q defined by q carried by the point $(1, 0, 0)$ to P. Now, given two points $A, B \in \mathbb{S}^2$, all rotations from A to B can be described. Take an arc $\overline{AB}$ on a great circle that joins both points. If R is a rotation that takes A to B, then the axis of R must be in the direction of the great circle that bisects $\overline{AB}$. Along that circle, there are two axes of rotation whose angle is easy to calculate:

1. When the axis of rotation passes through the midpoint M of $\overline{AB}$, the rotation angle θ is π radians. We will call that rotation R_1.
2. When the axis of rotation is perpendicular to the vectors $v = \vec{OA}$ and $w = \vec{OB}$, the rotation angle is $(\pm)$ the angle between v and w and is given by $\cos\theta = \langle v, w\rangle$. We will call this rotation R_2 (see Fig. B.3).

This reasoning allows us to calculate the axis and the angle of rotation for the two rotations that take $(1, 0, 0)$ to P. But we know that each axis and each angle of rotation defines a quaternion, hence we find q_1 and q_2 such $R_1 = R_{q_1}$ and $R_2 = R_{q_2}$. To give the explicit formulas for q_1 and q_2 we will proceed as follows. Given the point $P = (p_1, p_2, p_3)$, by the formulas for the angle and the axis of rotation, we

Fig. B.3 A rotation R_2

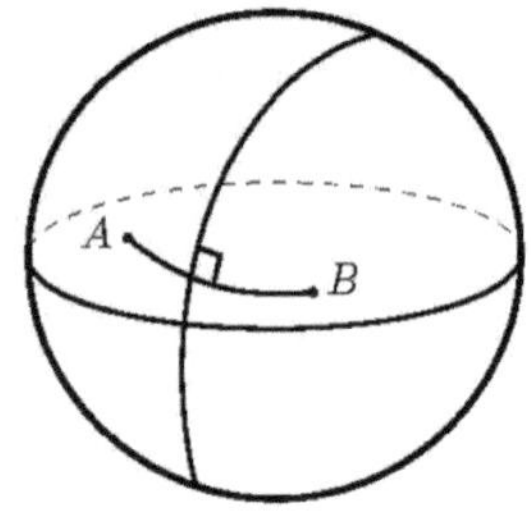

obtain

$$q_1 = \frac{1}{\sqrt{2(1+p_1)}}((1+p_1)i + p_2 j + p_3 k),$$

$$q_2 = \sqrt{\frac{1+p_1}{2}}(1 + \frac{-p_3}{1+p_1} j + \frac{p_2}{1+p_1} k).$$

So, writing $\cos t + i \sin t$ in the form e^{it}, the fiber $h^{-1}(P)$ is a circle in $\mathbb{R}^4$ parameterized from one of the following ways

$$h^{-1}(P) = \{q_1 e^{it}\}_{0\leq t\leq 2\pi} =$$

$$\{\frac{1}{\sqrt{2(1+p_1)}}(-(1+p_1)\sin t, (1+p_1)\cos t, p_3 \sin t + p_2 \cos t,$$
$$p_3 \cos t - p_2 \sin t)\}_{0\leq t\leq 2\pi};$$

$$h^{-1}(P) = \{q_2 e^{it}\}_{0\leq t\leq 2\pi}.$$

The point $(-1, 0, 0)$ is something special and we have

$$\begin{aligned} h^{-1}((-1,0,0)) &= \{k e^{it}\}_{0\leq t\leq 2\pi} \\ &= \{\cos tk + \sin tj\}_{0\leq t\leq 2\pi} \equiv \{(0,0,\sin t, \cos t)\}_{0\leq t\leq 2\pi}. \end{aligned}$$

Remember that the stereographic projection from the north pole $N = (0, 0, 1)$ is defined as

$$E_N : \mathbb{S}^2 - \{N\} \to \mathbb{R}^2,$$

$$(x, y, z) \mapsto (\frac{x}{1-z}, \frac{y}{1-z}),$$

where $\mathbb{R}^2$ has been identified with the plane $z = 0$. Its inverse is

$$(a, b) \mapsto (\frac{2a}{a^2+b^2+1}, \frac{2b}{a^2+b^2+1}, \frac{a^2+b^2-1}{a^2+b^2+1}).$$

It is easy to see that E_N transforms a circle of $\mathbb{S}^2$ containing N in a straight line, while in another case it becomes a plane circle (Fig. B.4).

This stereographic map is easily generalized to any dimension and, in particular, by choosing the point $(1, 0, 0, 0)$, we find out that

$$E : \mathbb{S}^3 - \{(1,0,0,0)\} \quad \to \quad \mathbb{R}^3,$$

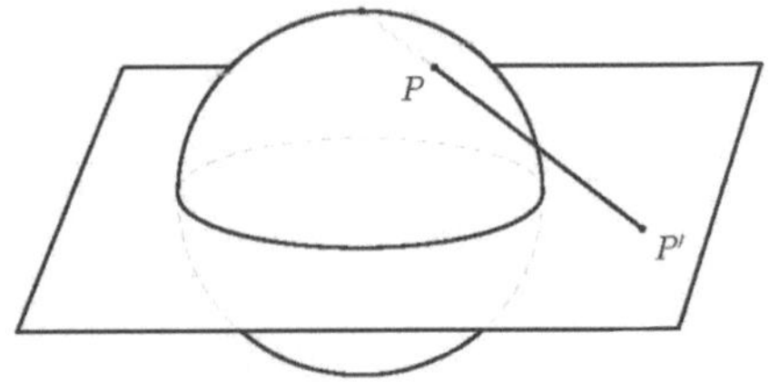

Fig. B.4 A stereographic projection

is given by

$$(w, x, y, z) \quad \mapsto \quad (\frac{x}{1-w}, \frac{y}{1-w}, \frac{x}{1-w}),$$

where $\mathbb{R}^3$ has been identified with the plane $z = 0$. Its inverse is

$$(a, b, c) \mapsto (\frac{a^2+b^2+c^2-1}{a^2+b^2+c^2+1}, \frac{2a}{a^2+b^2+c^2+1}, \frac{2b}{a^2+b^2+c^2+1}, \frac{2c}{a^2+b^2+c^2+1}).$$

We have seen that the fibers of the Hopf map are circles of $\mathbb{S}^3$, so its images by E are circles in $\mathbb{R}^3$, unless the fiber passes through the point $(1, 0, 0, 0)$. Then, through the composition

$$\mathbb{S}^2 \xrightarrow{h^{-1}} \mathbb{S}^3 - \{(1, 0, 0, 0)\} \xrightarrow{E} \mathbb{R}^3,$$

we see that $E \circ h^{-1}((1, 0, 0))$ is the x-axis, $E \circ h^{-1}((-1, 0, 0))$ is the unit circle in the plane y, z. For any other point $P = (p_1, p_2, p_3) \in \mathbb{S}^2$, other than $(1, 0, 0)$ and $(-1, 0, 0)$, $E \circ h^{-1}(P)$ is the circle in $\mathbb{R}^3$ given by

$$\frac{1}{(1+p_1)\sin t + \sqrt{2(1+p_1)})} \times ((1+p_1)\cos t, p_3 \sin t + p_2 \cos t, p_3 \cos t - p_2 \sin t),$$

that intersects the plane y, z exactly at two points A and B, one inside and one outside the unit circle in that plane. Furthermore, the plane containing the circle $E \circ h^{-1}(P)$ cannot contain the x-axis. Otherwise, it would cut the circle $E \circ h^{-1}((1, 0, 0))$, which cannot hold, since the fibers are disjoint. Therefore, the x-axis traverses the circle $E \circ h^{-1}(P)$, as outlined in the following figure (Fig. B.5)

As a summary, and because of the fact that "an image is worth more than thousand words", the Hopf map and its vision in $\mathbb{R}^3$ remain as follows (Fig. B.6):

Remark B.3.1 It should be noted that the version offered of the Hopf map is one of the many equivalents that appear in the literature and which does not even coincide

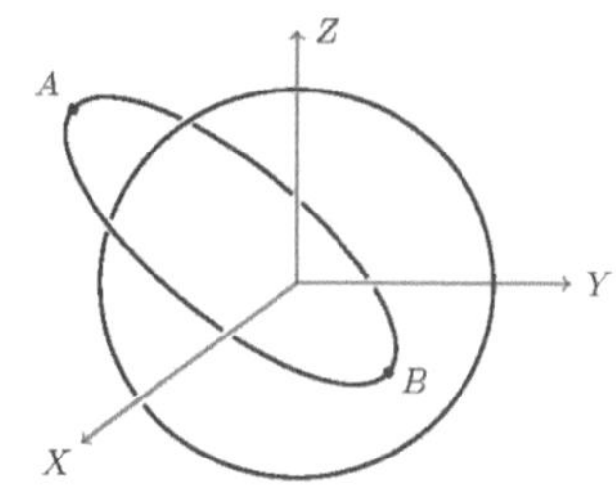

Fig. B.5 Stereographic projection of the fibers of the Hopf map

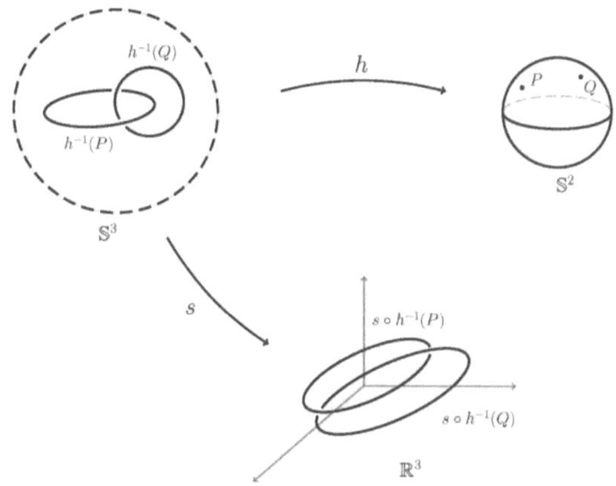

Fig. B.6 The Hopf map viewed in $\mathbb{R}^3$

with the one initially proposed by H. Hopf. In what follows, and to facilitate the calculations, we will use a different but equivalent formulation.

B.4 Third Definition of the Hopf Map

In many occasions it is more convenient to use complex coordinates complex, which can be done since $\mathbb{S}^2$ is the compactification by a point of the complex plane, that is, $\mathbb{S}^2 = \mathbb{C} \cup \{\infty\}$. The coordinate transformation is carried out through the stereographic projection. We will apply this idea to give a third definition of the Hopf map.

So, thinking about the stereographic projection $E : \mathbb{S}^2 - \{(1, 0, 0)\} \to \mathbb{R}^2$ from the point $Q = (1, 0, 0)$, take $P = (x, y, z) \in \mathbb{S}^2$ and its stereographic image $P' = E(P)$ in $\mathbb{R}^2$. As we know that P is in the segment $\overline{QP'}$, its coordinates are determined by the equation $P = (1 - t)Q + tP'$, $t \in [0, 1]$ (Fig. B.7).

As $\mathbb{R}^2$ has been identified with the plane $x = 0$, we will write $P' = (0, a, b)$, where $a = \mathrm{Re}(w)$ and $b = \mathrm{Im}(w)$ for a certain $w = a + i\,b \in \mathbb{C} \equiv \mathbb{R}^2$, that is, w will be the complex coordinate associated to the point P of the sphere. The equation of the segment leads to

$$x = 1 - t, \quad y = t\,a, \quad z = t\,b,$$

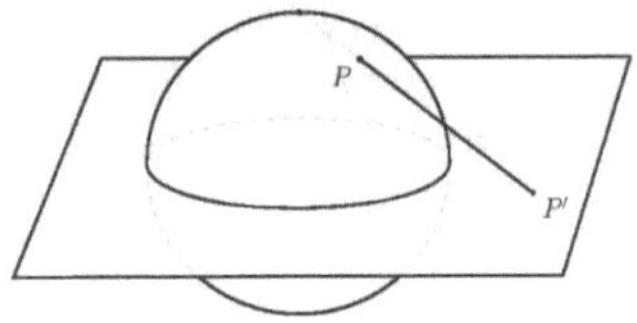

Fig. B.7 A stereographic projection

and the condition $x^2 + y^2 + z^2 = 1$ to $t = \dfrac{2}{w\,\bar{w} + 1}$, from which one deduces that

$$E^{-1} : w \mapsto \frac{1}{w\,\bar{w} + 1}(w\,\bar{w} - 1, 2\,\mathrm{Re}(w), 2\,\mathrm{Im}(w)),$$

i.e.,

$$x = \frac{w\,\bar{w} - 1}{w\,\bar{w} + 1}, \qquad y = \frac{2\,\mathrm{Re}(w)}{w\,\bar{w} + 1}, \qquad z = \frac{2\,\mathrm{Im}(w)}{w\,\bar{w} + 1}. \tag{B.3}$$

Obviously,

$$E : (x, y, z) \mapsto w = \frac{1}{1 - x}(y + i\,z),$$

and so, each point P of $\mathbb{S}^2$ can already be characterized either by a triple (x, y, z) of real numbers such that $x^2 + y^2 + z^2 = 1$, or by a complex number w related to the previous ones by the above equation.

In order to define the Hopf map, it is convenient to see the 3-sphere $\mathbb{S}^3 = \{(a, b, c, d) \in \mathbb{R}^4 : a^2 + b^2 + c^2 + d^2 = 1\}$ of the form

$$\mathbb{S}^3 = \{(w_1, w_2) \in \mathbb{C} : w_1\overline{w}_1 + w_2\overline{w}_2\}$$

where $w_1 = a + i\,b$ and $w_2 = c + i\,d$.

Definition B.4.1 (Third Definition) The Hopf map $h : \mathbb{S}^3 \to \mathbb{S}^2$ is defined as

$$h(w_1, w_2) = \frac{w_1}{w_2}.$$

Taking $\zeta = \dfrac{w_1}{w_2}$, a simple calculation from (B.3) leads to

$$x = w_1\overline{w}_1 - w_2\overline{w}_2, \qquad y + i\,z = 2w_1\overline{w}_2.$$

Therefore, the Hopf map is rewritten in the form

$$h(w_1, w_2) = (w_1\overline{w}_1 - w_2\overline{w}_2, 2\,\mathrm{Re}(w_1\overline{w}_2), 2\,\mathrm{Im}(w_1\overline{w}_2)) \tag{B.4}$$

$$= \left(a^2 + b^2 - c^2 - d^2, 2(ac + bd), 2(bc - ad)\right), \tag{B.5}$$

which, essentially, coincides with the previous definitions.

Given a unitary complex $w_o = e^{it},\ t \in [0, 2\pi]$, it is clear that

$$h(w_o w_1, w_o w_2) = h(w_1, w_2).$$

Therefore, $h^{-1}(a) = \mathbb{S}^1$, for each point $a \in \mathbb{S}^2$, that is, the fibers of Hopf map are great circles of $\mathbb{S}^3$. Also, given two distinct points $a, b \in \mathbb{S}^2$, the fibers $h^{-1}(a)$ and $h^{-1}(b)$ are linked. For example, if $a = (1, 0, 0)$ and $b = (-1, 0, 0)$, from (B.4) we arrive at the relations $w_1\overline{w}_2 = 1/2$ for a, and $w_1\overline{w}_2 = -1/2$ for b, which are satisfied, respectively, for the two circles of $\mathbb{S}^3$ given by

$$w_1 = w_2 = \frac{e^{it}}{\sqrt{2}}, \quad w_1 = -w_2 = \frac{e^{it}}{\sqrt{2}}. \tag{B.6}$$

The stereographic projection E, once again, helps us to see how they are linked. As we saw, we chose the point $Q = (1, 0, 0, 0)$ as the projection pole and set

$$P' = (x, y, z) = E(u_1, u_2, u_3, u_4) = \frac{1}{1 - u_1}(u_2, u_3, u_4).$$

So, the curves in $\mathbb{S}^3$ defined by (B.6) give us the following curves in $\mathbb{R}^3$ corresponding to $h^{-1}(a)$ and $h^{-1}(b)$, respectively,

$$C_a(t) = \frac{1}{\sqrt{2} - \cos t}(\sin t, \cos t, \sin t),$$

$$C_b(t) = \frac{1}{\sqrt{2} + \cos t}(-\sin t, \cos t, \sin t),$$

which are linked but do not intersect each other (Fig. B.8).

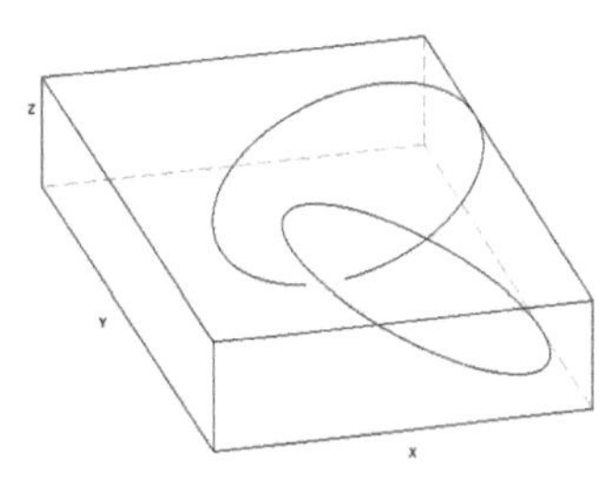

Fig. B.8 Two linked fibers

References

1. Alías, L.J., Ferrández, A., Lucas, P., Meroño, M.A.: On the Gauss map of B-scrolls. Tsukuba J. Math. **22**(2), 371–377 (1998)
2. Arroyo, J., Barros, M., Garay, O.J.: Some examples of critical points for the total mean curvature functional. Proc. Edinburgh Math. Soc. **43**, 587–603 (2000)
3. Arroyo, J., Barros, M., Garay, O.J.: Holography and total charge. J. Geom. Phys. **41**, 65–72 (2002)
4. Arroyo, J., Barros, M., Garay, O.J.: Models of relativistic particle with curvature and torsion revisited. Gen. Rel. Grav. **36**, 1441–1451 (2004)
5. Balgetir, H., Bektas, M., Ergüt, M.: On a characterization of null helix. Bul. Inst. Math. Acad. Sinica **29**(1), 71–78 (2001)
6. Barros, M.: General helices and a theorem of Lancret. Proc. A. M. S. **125**, 1503–1509 (2009)
7. Barros, M., Ferrández, A.: A conformal variational approach for helices in nature. J. Math. Phys. **50**, 103529 (2009)
8. Barros, M., Romero, A.: Magnetic vortices, Europhys. Lett. **77**, 34002, 5pp. (2007)
9. Barros, M., Ferrández, A.: On the energy density of helical proteins. J. Math. Biol. **69**, 1801–1813 (2014)
10. Barros, M., Ferrández, A., Lucas, P., Meroño, M.A.: Hopf cylinders, B-scrolls and solitons of the Betchov-Da Rios equation in the 3-dimensional anti-De Sitter space. C. R. Acad. Sci. Paris **321**, 505–509 (1995)
11. Barros, M., Ferrández, A., Lucas, P., Meroño, M.A.: General helices in the three-dimensional Lorentzian space forms. Rocky Mount. J. Math. **31**(2), 373–388 (2001)
12. Barros, M., Cabrerizo, J.L., Fernández, M., Romero, A.: The Gauss-Landau-Hall problem on Riemannian surfaces. J. Math. Phys. **46**, 112905 (2005)
13. Barros, M., Ferrández, A., Javaloyes, M.A., Lucas, P.: Relativistic particles with rigidity and torsion in $D = 3$ spacetimes. Class. Quant. Grav. **22**, 489–513 (2005)
14. Barros, M., Cabrerizo, J.L., Fernández, M., Romero, A.: Magnetic vortex filament flows. J. Math. Phys. **48**, 082904 (2007)
15. Barros, M., Caballero, M., Ortega, M.: Rotational surfaces in $\mathbb{L}^3$ and solutions of the nonlinear sigma model. Commun. Math. Phys. **290**, 437–477 (2009)
16. Barros, M., Ferrández, A., Garay, O.J.: Extremal curves of the total curvature in homogeneous three spaces. J. Math. Anal. Appl. **431**, 342–364 (2015)
17. Barros, M., Ferrández, A., Garay, O.J.: Dynamics of charges and solitons. J. Geom. Phys. **125**, 12–22 (2018)
18. Barros, M., Ferrández, A., Lucas, P., Meroño, M.A.: Helicoidal filaments in the 3-sphere (unpublished)

A. Ferrandez Izquierdo, *Helices*, RSME Springer Series 16,
https://doi.org/10.1007/978-3-032-09663-0

19. Bejancu, A.: Lightlike curves in Lorentz manifolds. Publ. Math. Debrecen **44**, 145–155 (1994)
20. Bejancu, A., Ferrández, A., Lucas, P.: A new viewpoint on geometry of a lightlike hypersurface in a semi-Euclidean space. Saitama Math. J. **16**, 31–38 (1999)
21. Berger, M.: Geometry II. Universitext. Springer, Berlin (1996)
22. Besse, A.L.: Einstein Manifolds. Springer Verlag (1987)
23. Betchov, R.: On the curvature and torsion of an isolated vortex filament. J. Fluid Mech. **22**, 471–479 (1965)
24. Bonnor, W.B.: Null curves in a Minkowski space-time. Tensor **20**, 229–242 (1969)
25. Borman, S., Washington, C.: Tying up loose ends: new examples and applications of circular and knotted peptides and proteins are turning up. Sci. Technol. **82**, 40–42 (2004)
26. Cabrerizo, J.L., Fernández, M., Ortega, M.: Massless particles in 3-dimensional Lorentzian warped products. J. Math. Phys. **48**, 012901 (2007)
27. Cahill, K.: Helices in biomolecules. Phys. Rev. E **72**, 062901 (2005)
28. Calugareanu, G.: Sur les classes d'isotopie des noeuds tridimensionnels et leurs invariants. Czech. Math. J. **11**, 588–625 (1961)
29. Cañadas, M.A., Gutiérrez, M., Ortega, M.: Massless particles in generalized Robertson-Walker 4-spaces. Ann. Mat. Pura Appl. **21**, 461–473 (2013)
30. Castagnino, M.: Sulle formule di Frenet-Serre per le curve nulle di una V4 riemanniana a metrica iperbolica normale. Rendiconti Mat. Roma Ser. **5**(23), 438–461 (1964)
31. Castrillón López, M., Fernández Mateos, V., Muñoz Masqué, J.: Total curvature of curves in Riemannian manifolds. Differ. Geom. Appl. **28**, 140–147 (2010)
32. Chouaieb, N., Goriely, A., Maddocks, J.H.: Helices. Proc. Natl. Acad. Sci. **103**(25), 9398–9403 (2006)
33. Cook, T.A.: The Curves of Life (Constable & Co. Reprinted Dover, London, 1979)
34. Daniel, B.: Isometric immersions into 3-dimensional homogeneous manifolds. Comment. Math. Helv. **82**, 87–131 (2007)
35. Dillen, F., Kühnel, W.: Ruled Weingarten surfaces in Minkowski 3-space. Manuscript. Math. **98**, 307–320 (1999)
36. Drutua-Romaniuc, S.L., Munteanu, M.I.: Magnetic curves corresponding to Killing magnetic fields in $\mathbb{E}^3$. J. Math. Phys. **52**, 113506, 10pp. (2011)
37. Duggal, K.L., Bejancu, A.: Lightlike Submanifolds of Semi-Riemannian Manifolds and Applications. Mathematics and its Aplications, vol. 364. Kluwer Academic Publishers Group, Dordrecht (1996)
38. Efimov, N.V.: Nekotorye zadachi iz teorii prostranstvennykh krivykh. Usp. Mat. Nauk **2**, 193–194 (1947)
39. Espinar, J.M., de Oliveira, I.S.: Locally convex surfaces immersed in a Killing submersion. Bull. Braz. Math. Soc. New Ser. **44**, 155–171 (2013)
40. Fenchel, W.: Über Krümmung und Windung geschlossener Raumkurven. Math. Ann. **101**, 238–252 (1929)
41. Fenchel, W.: The differential geometry of closed space curves. Bull. Am. Math. Soc. **57**, 44–54 (1951)
42. Feoli, A., Nesterenko, V.V., Scarpetta, G.: Functionals linear in curvature and statistics of helical proteins. Nucl. Phys. B **705**, 577–592 (2005)
43. Ferrández, A., Giménez, A., Lucas, P.: Null helices in Lorentzian space forms. Int. J. Modern Phys. A **16**(30), 4845–4864 (2001)
44. Ferrández, A., Giménez, A., Lucas, P.: Null generalized helices in Lorentz-Minkowski spaces. J. Phys. A Math. Gen. **35**, 8243–8251 (2002)
45. Ferrández, A., Giménez, A., Lucas, P.: Geometrical particle models on 3D null curves. Phys. Lett. B **543**, 311–317 (2002)
46. Furs, A.N., Barkovsky, M.L.: Complex Maxwell groups in the description of evanescent photons. J. Phys. A **31**(14), 3241–3253 (1998)
47. Galloway, J.: Helical imperative: paradigm of growth, form and function. In: Encyclopedia of Life Sciences (ELS). Wiley, Chichester (2010)

48. Graves, L.K.: Codimension one isometric immersions between Lorentz spaces. Trans. Am. Math. Soc. **252**, 367–392 (1979)
49. Gray, A.: Modern Differential Geometry of Curves and Surfaces with Mathematica, 2nd edn. CRC Press, Boca Raton (1998)
50. Greub, W., Halperin, S., Vanstone, R.: Connections, Curvature and Cohomology, vol. 3. Academic Press, Cambridge (1972)
51. Guven, J., Huber, G., Valencia, D.M.: Terasaki spiral ramps in the rough endoplasmic reticulum. Phys. Rev. Lett. **113**, 188101 (2014)
52. Hartman, P., Niremberg, L.: On spherical image maps whose Jacobians do not change sign. Am. J. Math. **81**, 901–920 (1959)
53. Hasimoto, H.: Motion of a vortex filament and its relation to elastica. J. Phys. Soc. Japan **31**, 293–294 (1971)
54. Hasimoto, H.: A soliton on a vortex filament. J. Fluid Mech. **51**, 477–485 (1972)
55. Hayden, H.A.: Deformations of a curve, in a Riemannian n-space, which displace certain vectors parallelly at each point. Proc. Lond. Math. Soc. **32**, 321–336 (1931)
56. Hayden, H.A.: On a generalized helix in a Riemannian n-space. Proc. Lond. Math. Soc. **32**, 337–345 (1931)
57. Hopf, H.: Über die Abbildungen der dreidimensionalen Sphäre auf die Kugelfläche. Math. Ann. **104**, 637–665 (1931)
58. Hou, Z.H., Ji, F.: On Lorentzian surfaces with $H^2 = K$ in Minkowski 3-space. J. Math. Anal. Appl. **334**, 54–58 (2007)
59. Hughston, L.P., Shaw, W.T.: Real classical strings. Proc. Roy. Soc. London Ser. A **414**, 415–422 (1987)
60. Hughston, L.P., Shaw, W.T.: Classical strings in ten dimensions. Proc. Roy. Soc. Lond. Ser. A **414**, 423–431 (1987)
61. Hughston, L.P., Shaw, W.T.: Constraint-free analysis of relativistic strings. Class. Quatum Gravity **5**, 69–72 (1988)
62. Hughston, L.P., Shaw, W.T.: Spinor parametrizations of minimal surfaces. In The Mathematics of Surfaces, III, pp. 359–372. Oxford University Press, New York (1989)
63. Ivey, T.: Helices, Hasimoto surfaces and Bäcklund transformations. Canadian Math. Bull. **43**, 427–439 (2000)
64. Konstantis, P.: Three-dimensional homogeneous spaces and their application in General Relativity. Ph.D. Tübingen (2013)
65. Kuznetsov, Y.A., Plyushchay, M.S.: The model of the relativistic particle with curvature and torsion. Nucl. Phys. B **389**(1), 181–205 (1993)
66. Lancret, M.A.: Memoire sur les courbes a double courbure. Memoires Presentes a l'Institut **1**, 416–454 (1806)
67. Langer, J., Perline, R.: Poisson geometry of the filament equation. J. Nonlinear Sci. **1**, 71–93 (1991)
68. Langer, J., Perline, R.: Local geometric invariants of integrable evolution equations. J. Math. Phys. **35**, 1732–1737 (1994)
69. Langer, J., Singer, D.A.: The total squared curvature of closed curves. J. Differ. Geom. **20**, 1–22 (1984)
70. Langer, J., Singer, D.A.: Knotted elastic curves in $\mathbb{R}^3$. J. London Math. Soc. **30**, 512–520 (1984)
71. Loh, C.H.: Multiple scale analysis of the spirally coiled cochlea. J. Acoust. Soc. Am. **74**(1), 95–103 (1983)
72. Lyons, D.W.: An elementary introduction to the Hopf fibration. Math. Mag. **76**(2), 87–98 (2003)
73. Manzano, J.M.: On the classification of Killing submersions and their isometries. Pac. J. Math. **270**(2), 367–392 (2014)
74. Marsden, J., Ratiu, T.: Introduction to Mechanics and Symmetry. Springer, New York (1994)
75. Millman, R.S., Parker, G.D.: Elements of Differential Geometry. Prentice-Hall, Hoboken (1977)

76. Mosseri, R., Dandoloff, R.: Geometry of entangled states, Bloch spheres and Hopf fibrations. J. Phys. A **34**, 10243–10252 (2001)
77. Nakahara, M.: *Geometry, Topology and Physics*. Institute of Physics Publishing, Philadelphia (1990)
78. Nersessian, A., Ramos, E.: Massive spinning particles and the geometry of null curves. Phys. Lett. B **445**, 123–128 (1998)
79. Nersessian, A., Ramos, E.: A geometrical particle model for anyons. Mod. Phys. Lett. A **14**(29), 2033–2037 (1999)
80. Nersessian, A., Manvelyan, R., Müller-Kirsten, H.J.W.: Particle with torsion on 3d null curves. Nucl. Phys. B **88**, 381–384 (2000)
81. O'Neill, B.: The fundamental equations of a submersion. Michigan Math. J. **13**, 459–469 (1966)
82. O'Neill, B.: Semi-Riemannian Geometry. Academic Press, Cambridge (1983)
83. Pauling, L., Corey, R.B.: Two pleated-sheet configurations of polypeptide chains involving both CIS and trans amide groups. Proc. Natl. Acad. Sci. USA **39**, 247 (1953)
84. Pauling, L., Corey, R.B.: Two rippled-sheet configurations of polypeptide chains, and a note about the pleated sheets. Proc. Natl. Acad. Sci. USA **39**, 253 (1953)
85. Pauling, L., Corey, R.B., Branson, H.R.: The structure of proteins: two hydrogen-bonded helical configurations of the polypeptide chain. Proc. Natl. Acad. Sci. USA **37**, 205 (1951)
86. Pinkall, U.: Hopf tori in $\mathbb{S}^3$. Invent. Math. **81**, 379–386 (1985)
87. Plyushchay, M.S.: Massless particle with rigidity as a model for the description of bosons and fermions. Phys. Lett. B **243**, 383–388 (1990)
88. Plyushchay, M.S.: Relativistic particle with torsion, Majorana equation and fractional spin. Phys. Lett. B **262**(1), 71–78 (1991)
89. Plyushchay, M.S.: The model of the relativistic particle with torsion. Nucl. Phys. B **362**(1–2), 54–72 (1991)
90. Rasor, S.E.: The geodesic lines on the helicoid. Ann. Math. **11**, 77–85 (1910)
91. Rios, L.D.: On the motion of an unbouded fluid with a vortex filament of any shape. Rend. Circ. Mat. Palermo **22**, 117–132 (1906)
92. Scott, P.: The geometries of three manifolds. Bull. London Math. Soc. **15**, 401–487 (1983)
93. Shaw, W.T.: Twistors and strings. In: Mathematics and General Relativity (Santa Cruz, CA, 1986), pp. 337–363. American Mathematical Society, Providence (1988)
94. Struik, D.J.: Lectures on Classical Differential Geometry. Dover, New York (1988)
95. Synge, J.L.: Geometry of dynamical null lines. Tensor **24**, 69–74 (1972)
96. Syptak, M.: Sur les hypercirconferences et hyperhelices dans les espaces euclidiens a p dimensions. C. R. Acad. Sci. Paris **195**, 298–299 (1932)
97. Syptak, M.: Sur les hypercirconferences et hyperhelices generalisees dans les espaces euclidiens a p dimensions. C. R. Acad. Sci. Paris **198**, 1665–1667 (1934)
98. Terasaki, M., et al.: Stacked endoplasmic reticulum sheets are connected by helicoidal membrane motifs. Cell **154**(2), 285–296 (2013)
99. Thurston, W.P.: Three dimensional manifolds, Kleinian groups and hyperbolic geometry. Bull. Am. Math. Soc. **6**, 357–381 (1982)
100. Thurston, W.P.: Three dimensional Geometry and Topology. Princeton University Press, New Jersey (1991)
101. Thompson, D.: On Growth and Form. Cambridge University Press, Cambridge (1917)
102. Urbantke, H.: On Pinl's representation of null curves in n dimensions. In: Relativity Today (Budapest, 1987), pp. 34–36. World Scientific Publishing, Teaneck (1988)
103. Urbantke, H.: The Hopf fibration -seven times in physics. J. Geom. Phys. **46**, 125–150 (2003)
104. Vessiot, E.: Sur les curbes minima. Comptes Rendus **140**, 1381–1384 (1905)
105. Watson, J.D., Crick, F.H.C.: Molecular structure of nucleids acids. Nature **171**, 737–738 (1953)
106. White, J.: Self-linking and the Gauss integral in higher dimensions. Am. J. Math. **91**, 693–728 (1969)
107. Wong, Y.C.: Generalized helices in an ordinary v_n. Proc. Camb. Philos. Soc. **37**, 14–28 (1941)

108. Wong, Y.C.: On the generalized helices of Hayden and Syptak in an N-space. Proc. Camb. Philos. Soc. **37**, 229–243 (1941)
109. Whitney, H.: On regular closed curves on the plane. Compos. Math. **4**, 276–284 (1937)
110. Zulli, L.: Charting the 3-sphere–An exposition for undergraduates. Am. Math. Month. **103**(3), 221–229

Terminology Index

A. Ferrandez Izquierdo, *Helices*, RSME Springer Series 16,
https://doi.org/10.1007/978-3-032-09663-0

FSC
www.fsc.org
MIX
Papier aus verantwortungsvollen Quellen
Paper from responsible sources
FSC® C105338

If you have any concerns about our products,
you can contact us on
ProductSafety@springernature.com

In case Publisher is established outside the EU,
the EU authorized representative is:
Springer Nature Customer Service Center GmbH
Europaplatz 3, 69115 Heidelberg, Germany

Printed by Libri Plureos GmbH
in Hamburg, Germany